Electrometallurgical Techniques for DOE Spent Fuel Treatment

FINAL REPORT

Committee on Electrometallurgical Techniques
for DOE Spent Fuel Treatment

Board on Chemical Sciences and Technology
Commission on Physical Sciences, Mathematics, and Applications
National Research Council

NATIONAL ACADEMY PRESS
Washington, D.C.

NOTICE: The project that is the subject of this report was approved by the Governing Board of the National Research Council, whose members are drawn from the councils of the National Academy of Sciences, the National Academy of Engineering, and the Institute of Medicine. The members of the committee responsible for the report were chosen for their special competences and with regard for appropriate balance.

Support for this project was provided by the National Academy of Sciences, the Howard Hughes Medical Foundation, the American Chemical Society, the U.S. Department of Energy, the National Institute of Standards and Technology, and the Camille and Henry Dreyfus Foundation. Any opinions, findings, conclusions, or recommendations are those of the author(s) and do not necessarily reflect the views of the organizations or agencies that provided support for this project.

International Standard Book Number 0-309-07095-3

Copyright 2000 by the National Academy of Sciences. All rights reserved.

Additional copies of this report are available from:
Board on Chemical Sciences and Technology
National Research Council
2101 Constitution Avenue, NW
Washington, D.C. 20418

Printed in the United States of America

THE NATIONAL ACADEMIES

National Academy of Sciences
National Academy of Engineering
Institute of Medicine
National Research Council

The **National Academy of Sciences** is a private, nonprofit, self-perpetuating society of distinguished scholars engaged in scientific and engineering research, dedicated to the furtherance of science and technology and to their use for the general welfare. Upon the authority of the charter granted to it by the Congress in 1863, the Academy has a mandate that requires it to advise the federal government on scientific and technical matters. Dr. Bruce M. Alberts is president of the National Academy of Sciences.

The **National Academy of Engineering** was established in 1964, under the charter of the National Academy of Sciences, as a parallel organization of outstanding engineers. It is autonomous in its administration and in the selection of its members, sharing with the National Academy of Sciences the responsibility for advising the federal government. The National Academy of Engineering also sponsors engineering programs aimed at meeting national needs, encourages education and research, and recognizes the superior achievements of engineers. Dr. William A. Wulf is president of the National Academy of Engineering.

The **Institute of Medicine** was established in 1970 by the National Academy of Sciences to secure the services of eminent members of appropriate professions in the examination of policy matters pertaining to the health of the public. The Institute acts under the responsibility given to the National Academy of Sciences by its congressional charter to be an adviser to the federal government and, upon its own initiative, to identify issues of medical care, research, and education. Dr. Kenneth I. Shine is president of the Institute of Medicine.

The **National Research Council** was organized by the National Academy of Sciences in 1916 to associate the broad community of science and technology with the Academy's purposes of furthering knowledge and advising the federal government. Functioning in accordance with general policies determined by the Academy, the Council has become the principal operating agency of both the National Academy of Sciences and the National Academy of Engineering in providing services to the government, the public, and the scientific and engineering communities. The Council is administered jointly by both Academies and the Institute of Medicine. Dr. Bruce M. Alberts and Dr. William A. Wulf are chairman and vice chairman, respectively, of the National Research Council.

COMMITTEE ON ELECTROMETALLURGICAL TECHNIQUES FOR DOE SPENT FUEL TREATMENT

GREGORY R. CHOPPIN, Florida State University, *Chair*
MICHAEL J. APTED, MonitorSci, Inc.
PATRICIA A. BAISDEN, Lawrence Livermore National Laboratory
EDITH M. FLANIGEN, UOP (retired)
CHARLES L. HUSSEY, University of Mississippi
FLORIAN MANSFELD, University of Southern California
L. EUGENE MCNEESE, Oak Ridge National Laboratory
ROBERT A. OSTERYOUNG, North Carolina State University
PAUL G. SHEWMON, Ohio State University
RALPH E. WHITE, University of South Carolina

Staff

CHRISTOPHER K. MURPHY, Study Director
DOUGLAS J. RABER, Director, Board on Chemical Sciences and Technology

BOARD ON CHEMICAL SCIENCES AND TECHNOLOGY

JOHN L. ANDERSON, Carnegie Mellon University, *Co-chair*
LARRY E. OVERMAN, University of California at Irvine, *Co-chair*
BARBARA J. GARRISON, Pennsylvania State University
ALICE P. GAST, Stanford University
LOUIS C. GLASGOW, DuPont Fluoroproducts
KEITH E. GUBBINS, North Carolina State University
NANCY B. JACKSON, Sandia National Laboratories
JIRI JONAS, University of Illinois at Urbana-Champaign
GEORGE E. KELLER II, Union Carbide Company (retired)
RICHARD A. LERNER, Scripps Research Institute
GREGORY A. PETSKO, Brandeis University
WAYNE H. PITCHER, JR., Genencor International, Inc.
KENNETH N. RAYMOND, University of California at Berkeley
PAUL J. REIDER, Merck Research Laboratories
LYNN F. SCHNEEMEYER, Bell Laboratories
MARTIN B. SHERWIN, ChemVen Group, Inc.
JEFFREY J. SIIROLA, Eastman Chemical Company
CHRISTINE S. SLOANE, General Motors
PETER J. STANG, University of Utah
JOHN T. YATES, JR., University of Pittsburgh
STEVEN W. YATES, University of Kentucky

DOUGLAS J. RABER, Director
RUTH MCDIARMID, Senior Program Officer
CHRISTOPHER K. MURPHY, Program Officer
SYBIL A. PAIGE, Administrative Associate
MARIA P. JONES, Senior Project Assistant

COMMISSION ON PHYSICAL SCIENCES, MATHEMATICS, AND APPLICATIONS

PETER M. BANKS, Veridian Corporation/ERIM International, Inc., *Co-chair*
W. CARL LINEBERGER, University of Colorado, *Co-chair*
WILLIAM F. BALLHAUS, JR., Lockheed Martin Corporation
SHIRLEY CHIANG, University of California at Davis
MARSHALL H. COHEN, California Institute of Technology
RONALD G. DOUGLAS, Texas A&M University
SAMUEL H. FULLER, Analog Devices, Inc.
JERRY P. GOLLUB, Haverford College
MICHAEL F. GOODCHILD, University of California at Santa Barbara
MARTHA P. HAYNES, Cornell University
WESLEY T. HUNTRESS, JR., Carnegie Institution
CAROL M. JANTZEN, Savannah River Technology Center
PAUL G. KAMINSKI, Technovation, Inc.
KENNETH H. KELLER, University of Minnesota
JOHN R. KREICK, Sanders, a Lockheed Martin Company (retired)
MARSHA I. LESTER, University of Pennsylvania
DUSA M. MCDUFF, State University of New York at Stony Brook
JANET L. NORWOOD, Former U.S. Commissioner of Labor Statistics
M. ELISABETH PATÉ-CORNELL, Stanford University
NICHOLAS P. SAMIOS, Brookhaven National Laboratory
ROBERT J. SPINRAD, Xerox PARC (retired)

MYRON F. UMAN, Acting Executive Director

Preface

The Committee on Electrometallurgical Techniques for DOE Spent Fuel Treatment was appointed by the National Research Council in 1994 as a result of an initial request by the U.S. Department of Energy (DOE) for an independent evaluation of the technical viability of electrometallurgical processing technology proposed by Argonne National Laboratory (ANL) as a potential approach for the treatment of DOE spent nuclear fuel. The committee completes its technical evaluation with this, its tenth and final report. Evaluation of the scientific progress of ANL's electrometallurgical program has remained at the core of the committee's charge throughout its existence. Within this core mission, the committee has also responded to other, more focused requests by DOE to examine specific issues related to the potential use of electrometallurgical technology. This evaluation has led to ten reports covering all aspects of ANL's electrometallurgical demonstration project.

As the specific tasks undertaken by the committee have evolved over the last five years, the committee membership has evolved accordingly as the National Research Council (NRC) has enlisted the assistance of volunteers with appropriate expertise (dates of service are in parentheses), whose contributions to the work of the committee are acknowledged here:

FRED BASOLO, Northwestern University, *Chair* (1994-1995),
GREGORY R. CHOPPIN, Florida State University, *Chair* (1996-2000),
JOHN F. AHEARNE, Duke University and Sigma Xi, The Scientific Research Society (1996-1997),
MICHAEL J. APTED, MonitorSci, Inc. (1994-2000),
PATRICIA A. BAISDEN, Lawrence Livermore National Laboratory (1994-2000),
SOL BURSTEIN, Wisconsin Electric Power Co. (retired) (1994-1996),
EDITH M. FLANIGEN, UOP (retired) (1996-2000),
CHARLES L. HUSSEY, University of Mississippi (1998-2000),
BERNARD KEAR, Rutgers University (1998-1999),
ALFRED F. LACAMERA, ALCOA Technical Center (1994-1995),
FLORIAN MANSFELD, University of Southern California (1998-2000),
L. EUGENE MCNEESE, Oak Ridge National Laboratory (1994-2000),
LAWRENCE J. MULLINS, Los Alamos National Laboratory (retired) (1994-1995),
ROBERT A. OSTERYOUNG, North Carolina State University (1994-2000),
JOHN D. SHERMAN, UOP (1996-1997),

PAUL G. SHEWMON, The Ohio State University (1999-2000),
RALPH E. WHITE, University of South Carolina (1998-2000),
JOEL D. WILLIAMS, Los Alamos National Laboratory (1996-1999), and
RAYMOND G. WYMER, Oak Ridge National Laboratory (retired) (1994-1997).

This study was conducted under the auspices of the NRC's Board on Chemical Sciences and Technology with assistance provided by its staff. The committee also acknowledges this support.

Gregory R. Choppin, *Chair*
Committee on Electrometallurgical Techniques for DOE Spent Fuel Treatment

Acknowledgment of Reviewers

This report has been reviewed in draft form by individuals chosen for their diverse perspectives and technical expertise, in accordance with procedures approved by the National Research Council's (NRC's) Report Review Committee. The purpose of this independent review is to provide candid and critical comments that will assist the authors and the NRC in making its published report as sound as possible and to ensure that the report meets institutional standards for objectivity, evidence, and responsiveness to the study charge. The review comments and draft manuscript remain confidential to protect the integrity of the deliberative process. We wish to thank the following individuals for their participation in the review of this report:

John F. Ahearne, Duke University and Sigma Xi, The Scientific Research Society,
Robert Budnitz, Future Resources Associates, Inc.,
Robert L. Fleischer, Union College,
Harold Forsen, Bechtel Corporation,
Lloyd Heldt, Michigan Technological University,
J. Brent Hiskey, University of Arizona,
Royce W. Murray, University of North Carolina at Chapel Hill, and
Thomas Pigford, University of California at Berkeley.

Although the reviewers listed above have provided many constructive comments and suggestions, they were not asked to endorse the conclusions or recommendations, nor did they see the final draft of the report before its release. The review of this report was overseen by Norman Hackerman (The Robert A. Welch Foundation), appointed by the Commission on Physical Sciences, Mathematics, and Applications, and Robert Connick (University of California at Berkeley), appointed by the Report Review Committee, who were responsible for making certain that an independent examination of this report was carried out in accordance with institutional procedures and that all review comments were carefully considered. Responsibility for the final content of this report rests entirely with the authoring committee and the NRC.

Contents

EXECUTIVE SUMMARY		1
1	INTRODUCTION	11
2	BACKGROUND AND DEVELOPMENT OF ELECTROMETALLURGICAL TECHNOLOGY FOR THE TREATMENT OF SPENT NUCLEAR FUEL	17
3	THE ELECTROMETALLURGICAL PROCESS AT ARGONNE NATIONAL LABORATORY	25
4	WASTE STREAMS PRODUCED BY THE ELECTROMETALLURGICAL TECHNOLOGY PROCESS	45
5	POST-DEMONSTRATION ACTIVITIES	61
6	ELECTROMETALLURGICAL TECHNOLOGY DEMONSTRATION PROJECT SUCCESS CRITERIA	71

APPENDIXES

A	Committee Charge and Statements of Task	79
B	Meeting Summary, July 21-22, 1999	85
C	Meeting Summary, September 19-21, 1999	103
D	Recommendations and Selected Findings and Conclusions from Previous Reports of the Committee on Electrometallurgical Techniques for DOE Spent Fuel Treatment	107
E	Abbreviations and Acronyms	115

Executive Summary

INTRODUCTION

The Committee on Electrometallurgical Techniques for DOE Spent Fuel Treatment was formed to evaluate the technical viability of electrometallurgical technology as a method for treating U.S. Department of Energy (DOE)[1] spent nuclear fuel (SNF).[2] Over the course of the committee's operating life, this charge has remained constant. Within the framework of this overall charge, the scope of the committee's work—as defined by its statements of task—has evolved in response to further requests from DOE, as well as technical accomplishments and regulatory and legal considerations. As part of its task, the committee has provided periodic assessments of Argonne National Laboratory's (ANL's) R&D program on the electrometallurgical technology.

In 1995, ANL proposed the use of electrometallurgical technology for treatment of all spent nuclear fuel in the DOE inventory.[3] Treatment would convert the fuel to components suitable for waste disposal as well as separate out any material that might be of use in future DOE operations. Electrometallurgical technology was suggested as a means to produce the same waste forms for all of the spent fuels in the DOE inventory, thus providing substantial cost savings for qualification of these waste materials for disposal in a geologic repository.

Electrometallurgical technology (EMT) consists of electrorefining the reactor fuel in an electrochemical cell. The fuel, in metallic form, is selectively dissolved at the anode while nearly pure uranium metal is deposited at the cathode, leaving fission products, fuel cladding material, plutonium, and other transuranic elements partially at the anode and partially in the molten salt electrolyte. Thus the fuel is separated into three components: metallic uranium, a metallic waste form from the anode, and a highly radioactive salt mixture that subsequently can be converted to a ceramic waste form. A key step in the ANL's proposal was treatment of all the Experimental Breeder Reactor-II (EBR-II) spent fuel as a demonstration of the technology. As work progressed on the EBR-II spent fuel, the committee's technical evaluation of electrometallurgical technology became increasingly focused on the demonstration project—which provided the primary source of data on which the committee could base its assessments.

---

[1] Acronyms and abbreviations are defined in Appendix E.
[2] DOE spent nuclear fuel refers to such fuels accumulated within the DOE complex; commercial production fuels are not included.
[3] Argonne National Laboratory, *Proposal for Development of Electrometallurgical Technology for Treatment of DOE Spent Nuclear Fuel*, Argonne National Laboratory, Argonne, IL, 1995.

A total of approximately 2,000 metric tons of SNF—broadly classified as production fuels, special fuels, or naval fuels—has accumulated throughout the DOE complex.[4]

The EMT process developed by ANL and originally proposed for the treatment of all DOE SNF is potentially applicable to a fairly wide variety of spent fuel types besides the EBR-II used by Argonne in its development and demonstration of the technology. For example, Fermi-1 blanket fuel and Fast Flux Test Facility sodium-bonded fuel are SNFs that can potentially be treated using EMT. DOE initially proposed that the EBR-II driver fuel and at least half of the blanket fuel be treated via the process.[5] For most fuels, such as oxides, the fuel would first have to be converted to a suitable metallic form before the electrorefining could be applied.

Electrometallurgical technology for treatment of DOE spent fuel evolved from ANL's work on the Advanced Liquid-Metal Reactor Integral Fast Reactor (ALMR/IFR).[6] With the termination of the ALMR/IFR project, this process, with some modification, served as the basis for ANL's January 1995 proposal, which included the use of the electrometallurgical process for the treatment of EBR-II spent nuclear fuel. The proposal was accepted by DOE and was to include treatment of both reactor driver fuel and uranium blanket material. The present committee as part of its task was asked to evaluate the ongoing work on electrometallurgical technology at ANL.

During its second year of operation, the committee was asked to evaluate the scientific and technological issues associated with extending ANL's electrometallurgical program to handle plutonium, in the event that DOE might pursue an electrometallurgical treatment option for the disposition of excess weapons plutonium (WPu). The committee concluded that disposition of WPu would involve different feeds for use in SNF processing, raising several concerns with respect to electrometallurgical processing, zeolite loading, and waste form performance.

THE ELECTROMETALLURGICAL PROCESS AT ANL

The electrometallurgical treatment process for spent fuel at ANL consists of several distinct steps: chopping the fuel elements, electrorefining the driver and blanket fuel, removing entrained salt from uranium electrodeposits and consolidating dendritic deposits in a cathode processor, casting separately into ingots the uranium metal from the cathode and the metal residue from the anode, and, finally, mixing, heating, and pressing the salt electrolyte with zeolite to form a ceramic waste.

The electrorefining step is the heart of the EMT process. The fuel element choppers are pneumatic punch presses that have been modified with blades for shearing driver and blanket fuel elements into segments for loading into the anode compartments of the Mark-IV electrorefiner (for driver fuel) and the Mark-V electrorefiner (for blanket fuel) developed at ANL. As part of ANL's demonstration project criterion that required a blanket throughput rate of 150 kg per month sustained for 1 month,[7] the blanket element chopper was used to process 3.5 blanket fuel assemblies or 66 blanket fuel elements, for a total of 164.4 kg of uranium.

The Mark-IV electrorefiner has an overall anode batch size of 16 kg and was designed for processing driver fuel. The efficiency of the overall electrorefining operation is enhanced by using a second cathode inserted into melt through the fourth port. The cadmium pool in the bottom of the Mark-IV catches and dissolves any of the uranium deposits that either fall off the cathode or are scraped off the cathode by the scrapers during the electro-deposition process.

During the repeatability phase of the demonstration project, the Mark-IV electrorefiner was used to treat 12 driver assemblies at an average rate of 24 kg of uranium per month over a 3-month period compared to the target of 16 kg (~4 driver assemblies) per month over a 3-month period.[8]

---

[4]Argonne National Laboratory, *Proposal for Development of Electrometallurgical Technology for Treatment of DOE Spent Nuclear Fuel*, Argonne National Laboratory, Argonne, IL, 1995.

[5]Argonne National Laboratory, *Proposal for Development of Electrometallurgical Technology for Treatment of DOE Spent Nuclear Fuel*, Argonne National Laboratory, Idaho Falls, ID, 1995.

[6]National Research Council, *Nuclear Wastes: Technologies for Separations and Transmutation*, National Academy Press, Washington, D.C., 1996, pp. 27-28, 43, 155-158.

[7]The full success criteria for the demonstration project, along with the goals to meet them, are included in Chapter 6.

[8]R.D. Mariani, D. Vaden, B.R. Westphal, D.V. Laug, S.S. Cunningham, S.X. Li, T.A. Johnson, J.R. Krsul, and M.J. Lambregts, *Process Description for Driver Fuel Treatment Operations*, NT Technical Memorandum No. 111, Argonne National Laboratory, Argonne, IL, 1999.

The second electrorefiner, the Mark-V, was designed for processing EBR-II blanket fuel, which is present in larger quantities and with lower enrichment than driver fuel. The basic difference between the Mark-IV and Mark-V is the design of the anodes and cathodes, which allowed significant increase in throughput. The material that collects in the product collector basket consists of uranium and approximately 20 wt % salt. Each anode-cathode module (ACM) in the Mark-V is capable of producing about 87 to 100 kg of uranium per month. Over the course of a 30-day process repeatability operation, the Mark-V was able to process the equivalent of 4.3 blanket assemblies (206 kg uranium (U) per month).

The purpose of the cathode processor is twofold: to remove entrained salt (and any cadmium) from the uranium electrodeposits by evaporation and to consolidate dendritic deposits.

The casting furnace provides a means to reduce the ^{235}U enrichment of the driver fuel product from the cathode processor by the addition of depleted uranium and to further consolidate the uranium product. The operating parameters associated with the casting furnace include crucible coating, temperature control, and pressure control.

Fabrication of Waste Forms

Following the electrorefining operations the stainless-steel cladding hulls are left in the anode basket, along with the noble metal fission products, some actinides, and adhering salt electrolyte. The uranium content is about 4 wt %; zirconium (Zr) metal is added to improve performance properties and to produce a lower-melting-point alloy. The material in the anode basket is placed in the cathode processor and heated to 1100 °C to distill the salt. The charge from the cathode processor is placed in an yttrium oxide crucible, is melted at approximately 1600 °C in the casting furnace in an argon (Ar) atmosphere, and then is cooled in the crucible and cast into ingots. The ingot constitutes the metal waste form (MWF).

The ceramic waste form (CWF) has been developed to immobilize the active fission products (alkalis, alkaline earths, and rare earths) and transuranic elements of the electrolyte. The CWF is produced in a batch process by mixing and blending the waste salt, periodically removed from the electrorefiner, with zeolite 4A at 500 °C to occlude the waste-loaded salt within the cages of the zeolite crystal lattice. Salt-loaded zeolite is mixed with a borosilicate glass and consolidated at high temperature (850 to 900 °C) and pressure (14,500 to 25,000 psi) in a hot isostatic press (HIP) to make the final waste form.

Pressureless sintering was investigated at ANL and may provide advantages over the HIP process during fabrication by giving a safer and easier pathway to volumetric scale-up of waste form fabrication. Further investigation of this potential capability, including the CWF produced using this process, is needed and is being pursued by ANL.

Recommendation: Studies to compare the type, abundance, and radionuclide inventory of minor and trace phases between ceramic waste forms produced by pressureless sintering versus the HIP process should be given high priority in the post-demonstration phase.

Alternatives to Electrometallurgical Technology for Treatment of DOE SNF

In two of its reports the committee, as part of its fulfillment of its tasks, evaluated alternatives to electrometallurgical technology:[9,10] direct disposal, glass material oxidation and dissolution, melt and dilute, PUREX,

---

[9]National Research Council, *An Assessment of Continued R&D into an Electrometallurgical Approach for Treating DOE Spent Nuclear Fuel*, National Academy Press, Washington, D.C., 1995. In this report the committee considered spent fuel treatment alternatives to EMT within the context of all DOE SNF.

[10]National Research Council, *Electrometallurgical Techniques for DOE Spent Fuel Treatment: Spring 1998 Status Report on Argonne National Laboratory's R&D Activity*, National Academy Press, Washington, D.C., 1998. In this report the committee considered spent fuel treatment alternatives within the context of EBR-II SNF.

chloride volatility, and plasma arc. While each of these technologies had merit, only PUREX was considered sufficiently well advanced to be a reasonable alternative to electrometallurgical treatment of spent fuel. However, the committee noted in its seventh report that public concern regarding transport of EBR-II spent fuel to the PUREX facility at the Savannah River site (SRS), combined with the expected shutdown of the PUREX canyons at SRS, argues against this alternative.[11]

WASTE STREAMS PRODUCED BY THE EMT PROCESS

Waste Form Qualification

The DOE, through its Office of Civilian Radioactive Waste Management (DOE-RW) and in conjunction with the development of final waste acceptance criteria to be based on Environmental Protection Agency/U.S. Nuclear Regulatory Commission regulations, is assessing the viability of permanent disposal of SNF in a deep geologic repository at Yucca Mountain, Nevada.[12] The performance and compatibility of the ANL waste forms must be assessed within this system context of overall repository safety.

The committee found that ANL's waste qualification strategy is appropriately based on guidance provided in the memorandum of agreement (MOA) between DOE-RW and DOE's Office of Environmental Management (DOE-EM).

Waste Acceptance Product Specifications

Data collected during the demonstration project provide information supporting issuance of an environmental impact statement (EIS) regarding continued application of the EMT process to the remaining inventory of EBR-II spent fuel. Thus, ANL has oriented its current activities to provide evidence of successful compliance with demonstration criteria.[13] Preliminary testing and modeling of the performance of EMT waste forms under repository conditions were also initiated during the demonstration project.

ANL's waste acceptance product specifications (WAPS)[14,15] are patterned after the quality assurance protocols used for Defense Program High-Level Waste (DHLW) borosilicate glass.[16] The committee observes, however, that DHLW borosilicate glass has *not* received final qualification and acceptance for geologic disposal by DOE-RW.

The committee understands that DOE is preparing waste acceptance criteria, including guidance on long-term waste form performance testing and qualification. This new document may modify the actual waste-acceptance strategies and waste-acceptance criteria that ANL's EMT program is currently following. These final criteria will influence long-term testing of the EMT metal waste form and the ceramic waste form, HLW waste forms intended for final disposition in a geologic repository.

---

[11]National Research Council, *Electrometallurgical Techniques for DOE Spent Fuel Treatment: Spring 1998 Status Report on Argonne National Laboratory's R&D Activity*, National Academy Press, Washington, D.C., 1998, p. 20.

[12]Office of Civilian Radioactive Waste Management, *Viability Assessment of a Repository at Yucca Mountain*, DOE/RW-0508, Department of Energy, Washington, D.C., 1998.

[13]National Research Council, *Electrometallurgical Techniques for DOE Spent Fuel Treatment: Fall 1996 Status Report on Argonne National Laboratory's R&D Activity*, National Academy Press, Washington, D.C., 1997. Also see Chapter 6 of this report.

[14]T.P. O'Holleran, R.W. Benedict, and S.G.. Johnson, *Waste Form Qualification Strategy for the Metal and Ceramic Waste Forms from Electrometallurgical Treatment of Spent Nuclear Fuel*, NT Technical Memorandum No. 115, Argonne National Laboratory, Argonne, IL, 1999.

[15]T.P. O'Holleran, D.P. Abraham, J.P. Ackerman, K.M. Goff, S.G. Johnson, and D.D. Keiser, *Waste Acceptance Product Specifications for the Waste Forms from Electrometallurgical Treatment of Spent Nuclear Fuel*, NT Technical Memorandum No. 116, Argonne National Laboratory, Argonne, IL, 1999.

[16]Westinghouse Savannah River Company, *DWPF Waste Acceptance Reference Manual (U)*, WSRC-IM-93-45, Savannah River Site, Aiken, SC, 1993.

Metal Waste Forms

The MWF test plan consists of attribute tests, characterization tests, accelerated tests, and service condition tests. Good progress to date seems to have been achieved in the identification of the various phases of stainless steel-15 zirconium (SS-15Zr)-type materials.[17] The characterization tests that have been terminated showed either no corrosive attack or only minor tarnish. ANL plans a total of 856 tests as necessary to achieve the goals of the project. In the view of the committee this seems an excessive number. Final corrosion tests from the demonstration project showed corrosion rates that were very low, and no correlation of elemental leaching with alloy composition was found. Tests have not yet been completed with added uranium. Electrochemical corrosion testing data presented at the end of the demonstration showed corrosion rates of the MWF alloys in J-13 (simulated Yucca Mountain well water) and in solutions of pH = 2, 4, and 10 that were low and similar to those of SS316 and alloy C22.

Finding: Some of the corrosion products, which may sequester radionuclides, might remain on the sample surface and might therefore not be detected by solution analysis.

Recommendation: Surface analysis by X-ray photoelectron spectroscopy (XPS) or Auger electron spectroscopy (AES) should be continued in the post-demonstration phase for selected samples to determine the chemical composition of passivating films and/or corrosion products.

According to ANL personnel in presentations to the committee,[18] galvanic corrosion tests according to ASTM G71 have indicated that enhanced corrosion of SS-15Zr due to galvanic coupling of the MWF with the inner lining of the waste form container (assumed to be alloy C22) is not likely to be significant. Vapor hydration tests found that corrosion rates were greatly accelerated by exposure to steam. The chemical nature of the corrosion products is under investigation. The effect of radiation on corrosion behavior has been discussed only briefly in presentations to the committee by ANL personnel. The toxicity characteristic leaching procedure (TCLP) test data suggest that the MWF passes the TCLP test.[19]

Recommendation: In the post-demonstration phase, ANL personnel should subject a few carefully selected samples to additional evaluation by surface analysis to determine the chemical composition of the corrosion products.

Recommendation: ANL personnel should concentrate on a few key samples, expose them at higher temperatures and chloride concentrations, and obtain electrochemical and surface analysis data.

Waste form degradation/radionuclide release models have been established that are an integral part of ANL's waste form repository performance assessment effort and will be used for predicting the long-term corrosion behavior of the MWF.

Ceramic Waste Form

To support repository qualification of the CWF, ANL developed a protocol and conducted a variety of tests and analyses relevant to the WAPS requirements. Detailed results and conclusions are contained in ANL's

---

[17] Material balance estimates for the MWF are given in Table 4.1.
[18] Presentations by Daniel Abraham and Dennis D. Keiser, Jr., to the committee, ANL-W, July 21, 1999.
[19] Presentation by Dennis D. Keiser, Jr., to the committee, ANL-W, July 21, 1999.

Ceramic Waste Form Handbook.[20] The CWF is a multiphase, nonhomogeneous composite consisting of approximately 75% sodalite, 25% borosilicate glass, and up to 5% other minor phases, e.g., aluminosilicates, rare-earth silicates, oxides, and halite (NaCl).[21] The CWF repository qualification program is based on the adaptation of models and test protocols developed for DHLW borosilicate glass. Accelerated alpha damage testing was carried out on simulated CWF doped with 0.2 to 2.5 wt % ^{238}Pu or ^{239}Pu. Initial results from this ongoing study show no significant degradation of the waste after 6 months at relatively low doses.[22]

> **Recommendation:** The electrometallurgical technology program should continue to investigate and evaluate in the post-demonstration period whether the test protocols and conceptual models developed for monolithic single-phase borosilicate glass can adequately represent the behavior of the nonhomogeneous multiphase CWF.

A variety of tests that monitor corrosion behavior were conducted by ANL to achieve a basic understanding of the processes that control dissolution of the CWF. Dissolution tests on the CWF over a 6-month period indicated that the CWF dissolves at a rate equal to or less than that of reference high-level waste borosilicate glass. The minor component actinides and rare earths form phases separate from the sodalite and glass phases. The actinides occur as nano-size (colloidal) crystal inclusions associated with the glass or the glass/sodalite grain boundaries.

> **Finding:** It is possible that some of these colloidal-sized crystal inclusions may be leached from the grain boundaries and that some may become colloidal suspensions with mobility much greater than expected from their solubility.[23]

Several mechanical and physical properties of the CWF were determined: cracking factor, thermal stability, fracture toughness, and density. The mechanical and physical properties of the CWF are comparable to or better than those of borosilicate high-level waste glass. Good product consistency is achieved using the specified demonstration HIP process parameters.

> **Finding:** The physical and mechanical behavior of the CWF under repository conditions should be comparable to that of borosilicate high-level waste glass.

Waste form performance has been modeled at ANL to predict the environmental impact of ANL's ceramic and metal waste forms on the proposed repository at Yucca Mountain. The model must be refined and verified with experimental data.

The committee found that the demonstration project success criteria (listed with the specific goals to meet them in Chapter 6) regarding the CWF have been met, although it is recognized that further data collection and

---

[20]W.L. Ebert, D.W. Esh, S.M. Frank, K.M. Goff, M.C. Hash, S.G. Johnson, M.A. Lewis, L.R. Morss, T.L. Moschetti, T.P. O'Holleran, M.K. Richmann, W.P. Riley, Jr., L.J. Simpson, W. Sinkler, M.L. Stanley, C.D. Tatko, D.J. Wronkiewicz, J.P. Ackerman, K.A. Arbesman, K.J. Bateman, T.J. Battisti, D.G. Cummings, T. DiSanto, M.L. Gougar, K.L. Hirsche, S.E. Kaps, L. Leibowitz, J.S. Luo, M. Noy, H. Retzer, M.F. Simpson, D. Sun, A.R. Warren, and V.N. Zyryanov, *Ceramic Waste Form Handbook*, NT Technical Memorandum No. 119, Argonne National Laboratory, Argonne, IL, 1999.

[21]Material balance estimates for the CWF are given in Table 4.1.

[22]Presentation by S.G. Johnson and L.R. Morss to the committee, ANL-W, July 21, 1999.

[23]For a study on the potential impact of actinides on repository performance, see A.B. Kersting; D.W. Efurd, D.L. Finnegan, D.J. Rokop, D.K. Smith, and J.L. Thompson, "Migration of Plutonium in Ground Water at the Nevada Test," *Nature*, Vol. 397, 1999, pp. 56-59.

analysis must be carried out in the post-demonstration period to support a final decision on CWF acceptability for repository disposal.

Finding: The committee sees no significant barriers to successful demonstration of an acceptable CWF, although full testing will extend beyond the demonstration time frame.

Recovered Uranium Material

As part of the EMT process, uranium metal is recovered at the cathode in the electrorefiner. After separation from the electrolyte it is cast as uranium metal ingots. During the cathode processing step and/or the casting step, natural or depleted uranium (DU) is added to the highly enriched uranium (HEU) derived from the EBR-II driver fuel. The disposition options for recovered uranium material are constrained by several DOE programmatic decisions and environmental impact statements. The depleted uranium recovered from treatment of the EBR-II blanket fuel is currently limited to indefinite storage or disposal as a transuranic (TRU) waste.

Finding: The current alternatives for disposition of uranium recovered from EBR-II fuel by electrorefining are limited to indefinite storage or speculative schemes for disposal.

Recommendation: The DOE should evaluate and select among these existing options for the disposition of recovered uranium in a timely manner so that the overall impacts of the EMT approach can be assessed.

POST-DEMONSTRATION ACTIVITIES

If DOE chooses to use the EMT process to treat sodium-bonded SNFs in the DOE inventory, or any other spent fuels,[24] ANL must complete all the activities required to qualify both the metal and ceramic waste baseline forms for repository disposal. ANL-E must also provide ongoing technical support to operations at ANL-W, and ANL-W must complete the required facility modifications and qualify the new, larger-scale equipment needed to handle the increased volume of fuel. These constitute a minimum set of post-demonstration activities. Post-demonstration qualification testing of EMT-produced waste forms must focus on the long-term rate of dissolution of the waste-form matrix, formation of radioactive element solubility-limiting solids, and potential formation of radionuclide-bearing colloids.

Recommendation: In its post-demonstration activities, ANL should reevaluate the appropriateness and applicability of its overall model to address the dissolution behavior and the multiphase nature of the EMT waste forms, especially the CWF. Associated test protocols, including that for the current product consistency test (PCT), should also be reevaluated.

A previous NRC report[25] criticized leach rate as a measure of the long-term performance of waste forms under expected repository conditions. The post-demonstration evaluation of the long-term performance of EMT waste forms, especially the CWF, under repository conditions must address this aspect of solubility limits for radioelements.

---

[24]Office of Civilian Radioactive Waste Management, *A Roadmap for Developing Accelerator Transmutation of Waste (ATW) Technology: A Report to Congress*, DOE/RW-0519, U.S. Department of Energy, Washington, D.C., 1999.

[25]National Research Council, *A Study of the Isolation System for Geologic Disposal of Radioactive Waste*, National Academy Press, Washington, D.C., 1983.

Recommendation: In the post-demonstration period, ANL should supplement and refine its current ASTM-based test protocols for waste form dissolution with respect to the technical perspectives on the long-term performance of the waste forms in geologic repositories, as described in the NRC's 1983 report by the Waste Isolation System Panel (WISP).[26]

There is considerable concern regarding the potential for rapid migration of significant quantities of radionuclides, especially Pu, at Yucca Mountain via colloidal transport.[27] The potential for formation and transport of radionuclide-bearing colloids should be specifically addressed in post-demonstration analysis and evaluation of EMT waste forms.

The committee observes that there may be alternative, nontesting approaches to assessing the acceptability of EMT waste forms for geologic disposal and that the merits of these alternatives would have to be technically evaluated by the DOE and by other independent peer reviews.

Recommendation: The eventual DOE waste acceptance criteria for geologic disposal should take into account available technical assessments. These waste acceptance criteria should be independently reviewed.

Should DOE decide to treat the remaining sodium-bonded spent fuel inventory, continuing efforts would be required to increase the capacity of some process equipment and to modify the facilities at ANL.

Recommendation: If the DOE decides to treat the remaining sodium-bonded spent fuel inventory and the waste form qualification efforts are successful, the required equipment upgrades and facility modifications should be adequately funded to ensure that treatment can be completed in a reasonable time and at a reasonable cost.

The use of pressureless sintering to produce the ceramic waste form can offer distinct advantages over the baseline HIP process. The potential advantages include a higher throughput per square foot of process space, increased safety, and reduced costs.

Recommendation: If pressureless sintering were to be used in place of the HIP process to produce the EMT ceramic waste form, waste form qualification studies would have to be conducted to determine its suitability for producing a waste form intended for deposit in a geologic repository.

In the post-demonstration period, continued development of a high-throughput electrorefiner (HTER), particularly if it could be cost-effectively developed and implemented in a timely fashion, could offer the advantage of considerably reducing the time required to treat the remaining sodium-bonded fuel. There are at least two options for increasing throughput up through the electrorefiner step in the EMT process. The first is continued development and implementation of a HTER (e.g., the 25-inch HTER under development at ANL-E) with a uranium deposition rate significantly exceeding that of the current Mark-V design. The second option is to simply double the current electrorefiner deposition rate by adding a second Mark-V electrorefiner to the Ar cell at ANL-W.

---

[26]National Research Council, *A Study of the Isolation System for Geologic Disposal of Radioactive Wastes*, National Academy Press, Washington, D.C., 1983.

[27]Office of Civilian Radioactive Waste Management, *Viability Assessment of a Repository at Yucca Mountain*, DOE/RW-0508, U.S. Department of Energy, Washington, D.C., 1998.

Recommendation: Continued development of a HTER should be evaluated in the context of the cost and time required for its development and implementation relative to the cost reduction that could be achieved by increasing the electrorefiner throughput by adding a second Mark-V and completing the inventory operations in the shorter time period.

ANL-E is pursuing the development of a production-scale zeolite column to increase the loading of the electrorefiner salt to about 3 wt % plutonium by running the salt through a column composed of zeolite. The use of the zeolite column could provide enhanced extraction and immobilization of fission products and actinides relative to a batch process, but a number of significant technical challenges remain. The removal of water during the early stages of elution might prove to be an intractable problem that could prevent the successful development of a zeolite column compatible with the EMT process.

Finding: The volume of sodium-bonded spent fuel waste generated using the "throw away salt" option, where a portion of the plutonium and fission-product-contaminated salt is mixed directly with zeolite and glass particles for waste disposal, is such a small fraction of the total waste destined for geologic disposal that waste volume reduction resulting from the use of the zeolite column would not have a significant impact on the overall waste disposal problem.

Recommendation: Continued development of the zeolite column should not be considered a high priority unless a compelling argument can be made that its development and implementation would significantly reduce waste disposal costs or associated costs of EMT treatment of the DOE sodium-bonded spent fuel inventory.

For EMT to be used to treat oxide fuels, a head-end step is required to convert the oxide to metal. ANL-E has been pursuing the use of lithium metal as a reducing agent in molten LiCl salts to effect this conversion. The committee concluded that the state of development of the lithium reduction head-end treatment step is fairly mature, and if it were allowed to go to completion, the DOE would have an additional option for treating uranium oxide spent nuclear fuel.

Recommendation: If the DOE wants an additional option besides PUREX for treating uranium oxide spent nuclear fuel, it should seriously consider continued development and implementation of the lithium reduction step as a head-end process to EMT.

ELECTROMETALLURGICAL TECHNOLOGY DEMONSTRATION PROJECT SUCCESS CRITERIA

The criteria proposed by ANL in 1998 for the demonstration project were similar in scope to those recommended by the committee in 1995 but smaller in scale in order to conform to the revised environmental assessment. The four criteria address the process, the waste streams, and the safety of the electrometallurgical demonstration project.

Finding: The committee finds that ANL has met all of the criteria developed for judging the success of its electrometallurgical demonstration project.

Finding: The committee finds no technical barriers to the use of electrometallurgical technology to process the remainder of the EBR-II fuel.

The EBR-II demonstration project has shown that the electrometallurgical technique can be used to treat sodium-bonded spent nuclear fuel. The major hurdle that remains is qualification of the waste forms from this processing. The total quantity of EBR-II spent nuclear fuel is relatively small, particularly in comparison to the total DOE spent fuel inventory, so even if qualification of the waste form were to prove impossible, the quantity of these materials that had been produced would be modest. The committee has found no significant technical barriers to the use of electrometallurgical technology to treat EBR-II spent fuel, and EMT therefore represents a potentially viable technology for DOE spent nuclear fuel treatment. However, before using EMT for processing other spent fuels in the DOE inventory, which would generate much larger amounts of these wastes than were produced in ANL's demonstration project, it would be necessary for these waste forms to receive the acceptance qualification.

1

Introduction

The Committee on Electrometallurgical Techniques for DOE Spent Fuel Treatment was formed in September 1994 in response to a request made to the National Research Council (NRC)[1] by the U.S. Department of Energy (DOE). DOE requested an evaluation of electrometallurgical processing technology proposed by Argonne National Laboratory (ANL) for the treatment of DOE spent nuclear fuel (SNF).[2] Electrometallurgical treatment of spent reactor fuel involves a set of operations designed to remove the remaining uranium metal and to incorporate the radioactive nuclides into well defined and reproducible waste streams.

Over the course of the committee's operating life, this charge has remained constant. Within the framework of this overall charge, the scope of the committee's work—as defined by its statement of task—has evolved in response to further requests from DOE, as well as technical accomplishments and regulatory and legal considerations. As part of its task, the committee has provided periodic assessments of ANL's R&D program on the electrometallurgical technology.

Electrometallurgical technology (EMT) consists of electrorefining the reactor fuel in an electrochemical cell. The fuel, in metallic form, is selectively dissolved at the anode while nearly pure uranium metal is deposited at the cathode, leaving fission products, fuel cladding material, plutonium and other transuranic elements partially at the anode and partially in the molten salt electrolyte. Thus the fuel is separated into three components; metallic uranium, a metallic waste form from the anode, and a highly radioactive salt mixture that subsequently can be converted to a ceramic waste form. In 1995, ANL proposed the use of electrometallurgical technology for treatment of all spent nuclear fuel in the DOE inventory.[3] Treatment would convert the fuel to components suitable for waste disposal as well as separate out any material that might be of use in future DOE operations. Electrometallurgical technology was suggested to offer the potential that it could produce the same waste forms for all of spent fuels in the DOE inventory, thus providing substantial cost savings for qualification of these wastes for disposal in a geologic repository. The ANL proposal was further based on the presumption that some form of treatment would be needed for part of the SNF inventory because the significant reactivity of certain spent fuels

---

[1]Acronyms and abbreviations are defined in Appendix E.
[2]DOE spent nuclear fuel (SNF) refers to such fuels accumulated within the DOE complex; commercial production fuels are not included.
[3]Argonne National Laboratory, *Proposal for Development of Electrometallurgical Technology for Treatment of DOE Spent Nuclear Fuel*, Argonne, IL, January 1995.

would preclude direct disposal in a geologic repository. Of particular concern was sodium-bonded fuel such as that from the EBR-II reactor at ANL's site in Idaho.

A key step in the ANL's proposal was treatment of all the EBR-II spent fuel as a demonstration of the technology. However, in response to a lawsuit, DOE scaled down the size of the demonstration so that it could be carried out under the scope of the existing Environmental Impact Statement (EIS).[4] As a result, the demonstration project was refocused to treat only part of the EBR-II spent fuel. As work progressed on the EBR-II spent fuel, the committee's technical evaluation of electrometallurgical technology became increasingly focused on the demonstration project—which provided the primary source of data on which the committee could make its assessments.

The committee, throughout its three phases of operation, has consistently recommended that ANL adopt a set of criteria by which the success of the use of electrometallurgical technology could be judged.[5,6] For the demonstration project, ANL adopted a set of four success criteria, together with specific technical goals to meet each of the criteria.[7] These success criteria are discussed in greater detail in Chapter 6.

During its tenure, the committee has operated under three different statements of task. Although in each phase it addressed a different aspect of the use of EMT for the treatment of SNF, the committee in all three phases operated under the general charge of evaluating the technical viability of electrometallurgical technology for treatment of DOE spent nuclear fuel.[8] This technical evaluation was performed based on the R&D pursued by ANL in its EBR-II electrometallurgical demonstration project.

THE COMMITTEE'S WORK

Phase One

In phase one, the committee evaluated ANL's research related to EMT that had been performed up to that point (i.e., the Integral Fast Reactor Program, see Chapter 2). In addition, the committee received briefings from experts in EMT, which provided a basis for the committee's assessment of ANL's R&D plan. Committee members also were briefed on ANL's decision to utilize EBR-II spent nuclear fuel for the EMT demonstration program. In phase one of its work, the committee did not address whether EMT should be adopted as a component of the national strategy for handling nuclear materials. Although it evaluated EMT in light of other technical options, the committee refrained at this stage from suggesting which option should be pursued.

The committee in phase one produced two reports on ANL's program plan for electrometallurgical technology:

- *A Preliminary Assessment of the Promise of Continued R&D into an Electrometallurgical Approach for Treating DOE Spent Fuel* (1995, Report 1),[9] and
- *An Assessment of Continued R&D into an Electrometallurgical Approach for Treating DOE Spent Nuclear Fuel* (1992, Report 2).

The committee recommended that ANL proceed with its development plan for an EMT demonstration.

---

[4]U.S. Department of Energy Office of Environmental Management Idaho Operations Office, *Department of Energy Programmatic Spent Nuclear Fuel Management and Idaho National Engineering Laboratory Environmental Restoration and Waste Management Programs Final Environmental Impact Statement*, DOE/EIS-0203-F, U.S. Department of Energy, Idaho Falls, ID, 1995.

[5]National Research Council, *An Assessment of Continued R&D into an Electrometallurgical Approach for Treating DOE Spent Nuclear Fuel*, National Academy Press, Washington, D.C., 1995, pp. S-7 and S-8.

[6]National Research Council, *Electrometallurgical Techniques for DOE Spent Fuel Treatment: Fall 1996 Status Report on Argonne National Laboratory's R&D Activity*, National Academy Press, Washington, D.C., 1997, p. 7.

[7]National Research Council, *Electrometallurgical Techniques for DOE Spent Fuel Treatment: Spring 1998 Status Report on Argonne National Laboratory's R&D Activity*, National Academy Press, Washington, D.C., 1998

[8]See Appendix A for the statements of task for the committee in each of the three phases of its work.

[9]Report numbers indicate the order of release of the nine committee reports issued to date by the National Research Council and published by the National Academy Press, Washington, D.C. The recommendations made in each of the reports are reproduced in Appendix D.

Phase Two

In July 1995, following a request by DOE, the NRC extended the tenure of its Committee on Electrometallurgical Techniques for DOE Spent Fuel Treatment. During this time, the EBR-II fuel treatment demonstration project had been delayed to allow for the completion of an environmental assessment (EA).[10] As a result of the EA, the amount of fuel in the demonstration was reduced and separation of plutonium during the processing was eliminated but the program was allowed to continue; any increase in scope would require that a full environmental impact statement be prepared.

This change in scope as a result of the EA had an impact on the committee's statement of task as it entered phase two (see Appendix A).

An objective of the committee was to learn how the modified scope, as defined by the EA, affected the adequacy of the demonstration and the potential application of the electrometallurgical technology to process DOE spent fuel.

In the second phase of its work the committee produced four reports: three status reports, and a fourth report on the issue of electrometallurgical treatment of excess weapons plutonium. This topic is discussed in detail in Chapter 2.

- *An Evaluation of the Electrometallurgical Approach for Treatment of Excess Weapons Plutonium* (1996, Report 3),
- *Electrometallurgical Techniques for DOE Spent Fuel Treatment: A Status Report on Argonne National Laboratory's R&D Activity* (1996, Report 4),
- *Electrometallurgical Techniques for DOE Spent Fuel Treatment: Fall 1996 Status Report on Argonne National Laboratory's R&D Activity* (1997, Report 5), and
- *Electrometallurgical Techniques for DOE Spent Fuel Treatment: Status Report on Argonne National Laboratory's R&D Activity Through Spring 1997* (1997, Report 6).

Phase Three

The committee entered its third phase of activity in 1998, focusing on specific aspects of the original charge (see Appendix A).[11]

The three reports issued in phase three all dealt with the scientific progress made in ANL's electrometallurgical demonstration project.

- *Electrometallurgical Techniques for DOE Spent Fuel Treatment: Spring 1998 Status Report on Argonne National Laboratory's R&D Activity* (1998, Report 7),
- *Electrometallurgical Techniques for DOE Spent Fuel Treatment: Status Report on Argonne National Laboratory's R&D Activity as of Fall 1998* (1999, Report 8), and
- *Electrometallurgical Techniques for DOE Spent Fuel Treatment: An Assessment of Waste Form Development and Characterization* (1999, Report 9).

In addition, in phase three the committee analyzed a number of other issues related to the demonstration project. In its Report 7 the committee evaluated a set of success criteria developed by ANL in response to a

---

[10]In response to concerns expressed by several nongovernmental organizations, DOE scaled back the scope of the demonstration project. A new environmental assessment (EA) was prepared, which stated that only part of the EBR-II spent fuel inventory would be treated in ANL's electrometallurgical demonstration project. Details of this EA are given in Chapter 6. Department of Energy, Science and Technology Office of Nuclear Energy, *Environmental Assessment: Electrometallurgical Treatment Research and Demonstration Project in the Fuel Conditioning Facility at Argonne National Laboratory – West*, DOE/EA-1148, Washington, D.C., 1996.

[11]Appendix A, in addition to the charge for phase three, also contains letters from DOE to the NRC clarifying this charge as it applies to EBR-II SNF.

previous recommendation made in Report 5 of the committee and judged the criteria as adequate for determining the success of the EBR-II Spent Nuclear Fuel Demonstration Project. These success criteria and the committee's evaluation of progress in meeting them are presented in Chapter 6.

In the same report the committee also evaluated electrometallurgical treatment of SNF in light of other possible treatment technologies. The committee's evaluation of these alternatives is reproduced in depth in Chapter 3.

In its ninth report, the committee examined the issue of waste forms produced by the electrometallurgical process. This topic was not explicitly part of the committee's statement of tasks for phase three but was addressed as a result of discussions between representatives from DOE and NRC. The ramifications of these waste forms are discussed in detail in Chapter 4, where the electrometallurgical waste streams are described.

This, the committee's tenth and final report, fulfills its tasks to assess the viability of electrometallurgical technology for treating DOE spent nuclear fuel and to monitor the scientific and technical progress of the ANL program on electrometallurgical technology, specifically within the context of ANL's demonstration project on electrometallurgical treatment of EBR-II SNF. Significant results are summarized and all recommendations of the first nine reports are reported in Appendix D. The electrometallurgical process, the equipment used, and the waste forms that result from this treatment are discussed and evaluated in detail. The committee also presents its evaluation of ANL's performance relative to the success criteria for the demonstration project, which have served as the basis for judging the efficacy of using electrometallurgical technology for the treatment of EBR-II spent nuclear fuel. Finally, this report addresses post-demonstration activities related to ANL's electrometallurgical demonstration project, and makes related recommendations in this area.

DISPOSITION OF SPENT NUCLEAR FUEL

DOE Inventory of Spent Nuclear Fuel

A total of approximately 2,000 metric tons of spent nuclear fuel has accumulated throughout the DOE complex.[12] The DOE has more than 150 different types of SNF stored at more than 200,000 units at DOE, non-DOE, and university facilities across the United States.[13] The fuels located at commercial nuclear reactors are currently believed to be suitable for direct disposal in a geologic repository.[14]

The SNF that is, or is scheduled to become, part of the DOE inventory is stored at nine DOE sites, eight miscellaneous facilities performing reactor and fuel development or testing and isotope generation, three "special case" commercial facilities, 33 U.S. universities, and 49 foreign research reactors.

SNFs are broadly classified as production fuels, special fuels, or naval fuels. Production fuels are located mainly at the DOE's Hanford Reservation and include the N-reactor fuels, which account for about 80% of the total DOE SNF inventory. Special fuels include both low- and high-enrichment fuels from a variety of reactors used in a wide range of research, development, and testing activities. Naval fuels are those developed and used for naval propulsion and for related R&D activities.

---

[12]Argonne National Laboratory, *Proposal for Development of Electrometallurgical Technology for Treatment of DOE Spent Nuclear Fuel*, Argonne National Laboratory, Argonne, IL, January 1995.

[13]Fillmore, Denzel L., and Kenneth D. Bulmahn, Characteristics of Department of Energy Spent Nuclear Fuel, in *Proceedings of the Topical Meeting on DOE Spent Nuclear Fuel*, Salt Lake City, Dec. 13-16, 1994, American Nuclear Society, LaGrange Park, IL, pp. 313ff. Unless otherwise noted, all data in this section are from this source.

[14]International Atomic Energy Agency, *International Fuel Cycle Evaluation—Summary Volume,* page 227, International Atomic Energy Agency, Vienna, Austria, 1980.

Electrometallurgical Techniques for Treatment of DOE Spent Fuels

Electrometallurgical technology was originally proposed by ANL as a process with the potential to successfully treat all DOE spent fuels. Among the earlier incentives to proceed with R&D on EMT was its potential for handling a variety of different spent fuels, such as N-reactor fuel from Hanford, Molten Salt Reactor Experiment (MSRE) residues, and Savannah River Site fuels. The EBR-II termination program employed EMT for the treatment of EBR-II driver fuel and blanket assemblies.[15] This was the basis for ANL's electrometallurgical demonstration project.[16] In its fourth report, the committee recommended that "upon satisfactory completion of the demonstration with EBR-II fuel, the electrometallurgical technique should be evaluated in the broader context of alternative technologies for processing spent nuclear fuel."[17]

The electrometallurgical technology developed by ANL is potentially applicable to a fairly wide variety of spent fuel types besides the EBR-II fuel used by Argonne in its development and demonstration of the technology, for example Fermi-1 blanket fuel or Fast Flux Test Facility sodium-bonded fuel. In principle, electrometallurgical separations in alkali metal chloride media could be used for most of the DOE SNF, if used in conjunction with appropriate head-end processes. For most fuels, such as oxides, the fuel would first have to be converted to a suitable metallic form before the electrorefining could be applied. The purpose of specially tailored head-end processes would be to convert the fuels to metal and free them from elements such as aluminum and carbon that, above certain threshold-level concentrations, are incompatible with the electrometallurgical process as currently operated.

DOE initially proposed that the EBR-II driver fuel and at least half of the blanket fuel be treated via the electrometallurgical process.[18] It is expected that, in addition to processing EBR-II fuel, the electrometallurgical technology could process undamaged metallic fuels from the Hanford N-reactor without any head-end treatment. N-reactor fuels constitute about 80% by mass of DOE's SNF inventory, and EBR-II fuels about 1%.

However, a significant fraction of the N-reactor fuel elements have a breach in their cladding, and these failures have resulted in oxidation of the uranium metal as well as the zirconium cladding. These failed fuel elements (and the accompanying sludge that has formed in Hanford's K-basin East) would have to be treated by head-end processing such as chemical purification and oxide reduction prior to electrorefining.

In addition to N-reactor fuel, Hanford has a wide variety of other SNFs stored on site. These fuels amount to only about 1.5% of the mass of N-reactor fuel, but the amount is not negligible (33 MT). About half of this amount by mass is the Shippingport pressurized water reactor Core II, treatment of which by the electrometallurgical process would require a special head-end step to remove a layer of graphite from the fuel wafers that contain the fissile material. Another (small) portion of the Hanford SNF contains aluminum that is metallurgically bonded to the uranium. It is unlikely that this fuel could be processed directly by the electrometallurgical process without additional head-end treatment to remove the aluminum. Aluminum's tendency to form low-melting eutectic mixtures and volatile species would present a significant challenge for containment and effluent treatment in the electrorefining process.

---

[15]Prior to its electrometallurgical demonstration project, ANL personnel had addressed a number of challenges in developing the electrometallurgical process. ANL personnel had demonstrated, on an engineering scale, electrometallurgical separations with simulated metal fuel. They had also demonstrated, on a laboratory scale, the lithium reduction of simulated mixed oxide (MOX) fuel.

[16]National Research Council, *An Assessment of Continued R&D into an Electrometallurgical Approach for Treating DOE Spent Nuclear Fuel*, National Academy Press, Washington D.C., 1995.

[17]National Research Council, *Electrometallurgical Techniques for DOE Spent Fuel Treatment: A Status Report on Argonne National Laboratory's R&D Activity*, National Academy Press, Washington, D.C., 1996, p. 2.

[18]Argonne National Laboratory, *Proposal for Development of Electrometallurgical Technology for Treatment of DOE Spent Nuclear Fuel*, Argonne National Laboratory, Idaho Falls, ID, 1995.

Argonne National Laboratory's Electrometallurgical Technology

The electrometallurgical technology under development at ANL is derived from many years of research and development on molten salt systems for the production of materials for nuclear reactors and weapons as well as from activities in battery development. This technology for treating DOE's SNF consists of several unit operations:

- Head-end treatment, including fuel disassembly and steps such as oxide reduction, if required;
- Electrorefining; and
- Treatment of effluent electrorefining streams, including the ceramic and metal waste forms produced by this electrorefining, and processing of the uranium deposited at the steel cathode.

It is the technological aspects of these operations that the Committee on Electrometallurgical Techniques for DOE Spent Fuel Treatment has evaluated in the course of its work from 1995 to the present.

2

Background and Development of Electrometallurgical Technology for the Treatment of Spent Nuclear Fuel

INTRODUCTION

The electrometallurgical technique for treatment of DOE spent fuel, and in particular its application to the EBR-II Spent Nuclear Fuel Treatment Demonstration Project conducted by Argonne National Laboratory from June 1996 through June 1999, evolved in large part from ANL's earlier work on the Advanced Liquid-Metal Reactor/Integral Fast Reactor (ALMR/IFR).[1] The process developed was aimed initially at recycling IFR (and perhaps spent oxide fuels from light water reactors) into new IFR fuels, which would contain substantial quantities of uranium, plutonium, other actinides, and long-lived fission products that could then be burned in the IFR.[2,3] A liquid cadmium cathode was to be used for separation of the bulk of the plutonium and the other transuranic elements (TRUs) from the bulk of the uranium, which was electrolytically deposited (separated) at a steel cathode.[4] With the termination of the ALMR/IFR project, this process, with some modification, served as the basis for a proposal from Argonne in January 1995 for five major task areas: (1) treatment of metallic spent fuels; (2) recovery and treatment of canister and storage basin sludge; (3) treatment of oxide spent fuels; (4) waste treatment processes; and (5) waste form production and qualification.[5] This included the use of the electrometallurgical process for the treatment of EBR-II SNF. The proposal was accepted by DOE and was to include treatment of both reactor driver fuel and uranium blanket material. The present committee as part of its task was asked to evaluate the ongoing work on electrometallurgical technology at ANL. Use of the proposed process for

---

[1]National Research Council, *Nuclear Wastes: Technologies for Separations and Transmutation*, National Academy Press, Washington, D.C., 1996, pp. 27-28, 43, 155-158.

[2]National Research Council, *An Evaluation of the Electrometallurgical Approach for Treatment of Excess Weapons Plutonium*, National Academy Press, Washington, D.C., 1996, p. 1.

[3]A block diagram for the pyroprocess proposed for the IFR is available in National Research Council, *Nuclear Wastes: Technologies for Separations and Transmutation*, National Academy Press, Washington, D.C., 1996, p. 157.

[4]National Research Council, *An Assessment of Continued R&D into an Electrometallurgical Approach for Treating DOE Spent Nuclear Fuel*, National Academy Press, Washington, D.C., 1995, p. 7.

[5]National Research Council, *Electrometallurgical Techniques for DOE Spent Fuel Treatment: A Preliminary Assessment of the Promise of Continued R&D into an Electrometallurgical Approach for Treating DOE Spent Fuel*, National Academy Press, Washington, D.C., 1995, Appendix 1.

oxide fuels would have required a separate front end step to convert the oxides into metal for use in the electrorefiner.[6] This step consisted of reduction of the metal oxides into metal by Li, and electrochemical regeneration of metallic Li from Li_2O in molten LiCl. The process has also been considered as a possible technology for disposing of excess Pu from the U.S. stockpile.[7]

Pyroprocessing, or molten salt electrochemical processing, has been in general use for many years for purification of materials, including plutonium.[8] It involves anodization (oxidation) of a metal into a molten salt electrolyte and then reduction at a cathode to yield a more (highly) purified form. The overall process technology is diagrammed in Figure 2.1. Differences between the generic process and the pyroprocess as carried out by ANL are described below. Subsequent chapters provide additional detail.

ELECTROMETALLURGICAL TREATMENT

In the EBR-II demonstration project for the treatment of EBR-II driver and blanket assemblies, 100 driver assemblies, consisting of 410 kg of 60 to 75% highly enriched ^{235}U, and up to 25 blanket assemblies, consisting of 1,200 kg of depleted uranium, were to be treated. (As originally proposed, the EBR-II demonstration project was to have treated 1 metric ton of driver fuel and 16 metric tons of blanket fuel, the latter in a high throughput electrorefiner.[9,10] The demonstration project was limited to the size noted as the result of a revised EA.[11])

As indicated in Figure 2.1, chopped driver (or blanket) fuel rod elements are placed in a steel anode basket in an electrorefiner that contains a KCl-LiCl molten salt eutectic system at upwards of 500 °C. The driver fuel is highly enriched uranium alloyed with ~10 wt % Zr; the cladding is stainless steel. The blanket fuel is depleted uranium, with stainless steel cladding. Both the driver and the blanket elements are sodium-bonded to the stainless steel. Two electrorefiners, the Mark-IV and Mark-V, were developed by ANL for use in the EBR-II demonstration project. The Mark-IV, used to electrorefine the driver elements from the EBR-II, contains a molten Cd pool—a holdover from the ALMR/IFR development—while the Mark-V, used for the blanket elements, is Cd free. The Cd pool provides a corrosion-resistant barrier to the mild steel vessel and also acts as a neutron absorber to prevent criticality problems that might result from the highly enriched uranium in the driver elements falling to the bottom of the vessel. However, the Cd pool in the Mark-IV electrorefiner is not used as a cathode, and thus no Pu separation is performed. The Mark-V differs markedly from the Mark-IV, in that it is designed to process much larger batches of material, as needed to treat the blanket elements; it is also a high-throughput electrorefiner (HTER). Nevertheless, the fundamentals of the two electrorefiners are very much the same.

An oxidant, either $CdCl_2$, in the case of the Mark-IV, or UCl_3, in the case of the Mark-V, is added to the salt prior to initiation of electrolysis. The $CdCl_2$ oxidizes some of the U (and other active metals) from the anode baskets. Upon passage of a constant electrolysis current between the anode baskets and the steel cathode, U, Pu, transuranic elements (TRU), the alkalis and alkaline earth metals, and rare earths are oxidized into the molten salt as U^{3+}, Pu^{3+}, TRU, alkali and alkaline earth and rare-earth cations (Table 2.1). The stainless steel from the cladding, most of the Zr, and the noble metals remain in the anode baskets. The U^{3+} is reduced to the metal and

---

[6]National Research Council, *Electrometallurgical Techniques for DOE Spent Fuel Treatment: A Preliminary Assessment of the Promise of Continued R&D into an Electrometallurgical Approach for Treating DOE Spent Fuel*, National Academy Press, Washington, D.C., 1995, p. 7.

[7]National Research Council, *An Evaluation of the Electrometallurgical Approach for Treatment of Excess Weapons Plutonium*, National Academy Press, Washington, D.C., 1996, p. 1.

[8]M.R. Coops, J.B. Knighton, and L.J. Mullins, *Plutonium Chemistry*, W.T. Carnall and G.R. Choppin, eds., ACS Symposium Series 216, American Chemical Society, Washington, D.C., 1983, pp. 381-400.

[9]Argonne National Laboratory, *Proposal for Development of Electrometallurgical Treatment for DOE Spent Fuel*, Argonne National Laboratory, Argonne, IL, 1995.

[10]National Research Council, *An Assessment of Continued R&D into an Electrometallurgical Approach for Treating DOE Spent Nuclear Fuel*, National Academy Press, Washington, D.C., 1995, p. 12.

[11]National Research Council, *Electrometallurgical Techniques for DOE Spent Fuel Treatment: A Status Report on Argonne National Laboratory R&D Activity*, National Academy Press, Washington, D.C., 1996, p. 5.

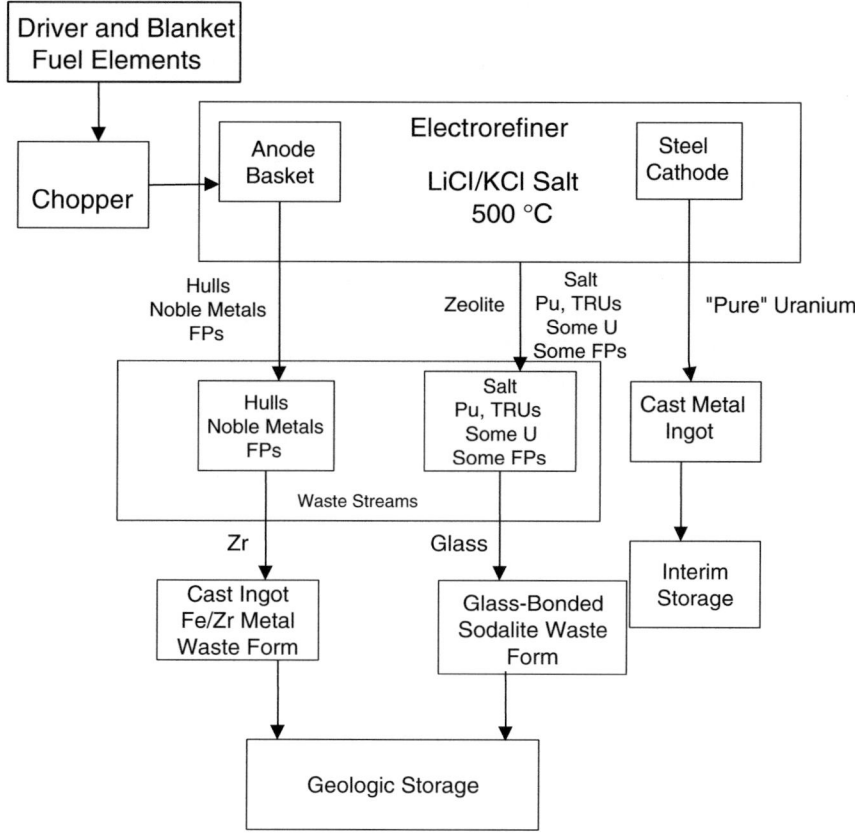

FIGURE 2.1 Overview block diagram of ANL's electrometallurgical process for DOE spent nuclear fuel. This is the process as originally proposed by ANL for the treatment of spent nuclear fuel.

TABLE 2.1 Free Energies of Formation of SNF Chlorides ($-\Delta G_o$ in units of kcal/g-eq at 500 °C)

Elements That Remain in Salt (very stable chlorides)		Elements That Can Be Electrotransported Efficiently		Elements That Remain As Metals (less stable chlorides)	
$BaCl_2$	87.9	$CmCl_3$	64.0	$CdCl_2$	32.3
$CsCl$	87.8	$PuCl_3$	62.4	$FeCl_2$	29.2
$RbCl$	87.0	$AmCl_3$	62.1	$NbCl_5$	26.7
KCl	86.7	$NpCl_3$	58.1	$MoCl_4$	16.8
$SrCl_2$	84.7	UCl_3	55.2	$TcCl_4$	11.0
$LiCl$	82.5	$ZrCl_4$	46.6	$RhCl_3$	10.0
$NaCl$	81.2			$PdCl_2$	9.0
$CaCl_2$	80.7			$RuCl_4$	6.0
$LaCl_3$	70.2				
$PrCl_3$	69.0				
$CeCl_3$	68.6				
$NdCl_3$	67.9				
YCl_3	65.1				

SOURCE: Reproduced from National Research Council, *An Assessment of Continued R&D into an Electrometallurgical Approach for Treating DOE Spent Nuclear Fuel*, National Academy Press, Washington, D.C., 1995, p. 9. Data supplied by James Laidler, Argonne National Laboratory, January 1995.

deposited onto the cathode in a reasonably pure state. In essence, the U is electrotransported from the chopped fuel (or blanket) elements in the anode basket to the cathode. The electrolysis is carried out under controlled current conditions such that principally U^{3+} is reduced at the cathode. For this to happen, a reasonably controlled amount of U^{3+} must be maintained in the melt. However, the build-up of sodium ion in the LiCl-KCl eutectic ultimately raises the melting point of the initial eutectic salt and requires removal of some of the salt and addition of fresh LiCl-KCl.

After a given period of electrolysis, the U cathode is removed from the Mark-IV electrorefiner, and the adherent molten salt is volatilized off in a vacuum furnace (cathode processor) and returned to the electrorefiner; the U is then cast into an ingot in a high-temperature furnace (casting furnace). (In the Mark-V the U deposits on a cathode from which it is scraped into a collection basket under the electrolysis cell). In the case of the U from the driver fuel, depleted uranium is added as well. Ultimately, the material remaining in the anode basket (the stainless steel hulls and any unoxidized material, the noble metals, and some fission products) is also subjected to treatment to volatilize off the adherent molten salts and is cast into an ingot in the casting furnace, yielding the metal waste form. The baseline metal waste form contains ~15 wt % Zr; this requires that anode basket hulls from the blanket processing have Zr added in the casting furnace. At an appropriate time, the salt containing the Pu, TRU elements, alkalis and alkaline earths, and some fission products is removed from the refiner, mixed with zeolite, heated to adsorb the salt into the zeolite, and then mixed with glass and hot isostatically pressed into a glass-bonded sodalite (GBS), the ceramic waste form.

Table 2.1 lists the free energies of formation per gram-equivalent of a number of the chlorides whose metals are important constituents of the spent nuclear fuel in the EBR-II. The elements are listed in three columns, going from those whose chlorides are most stable to those whose chlorides are least stable. On anodization of the SNF in the anode baskets, the metallic elements in the first two columns are, for the most part (the exception being Zr), converted into their chlorides and dissolve in the LiCl-KCl chloride eutectic. The metallic elements listed in the third column of Table 2.1 are sufficiently noble that they remain in the anode baskets in the metallic state. The electrolysis is carried out under a constant, controlled current. Limits are placed on the cell voltage, and thus, in effect, on both the anode and cathode potentials, such that

- Zr is essentially not anodized into the eutectic molten salt and
- Only the U^{+3} is reduced at the cathode.

If Zr were anodized into the salt, it would be reduced at the cathode. Thus, the TRUs in the first column of Table 2.1 are anodized into the molten salt but are not reduced at the cathode under the carefully controlled electrolysis conditions. As originally proposed by ANL, Pu and the bulk of the TRU elements would have been oxidized from the anode basket and deposited at the liquid Cd cathode in the Mark-IV electrorefiner (Figure 2.2); these elements form intermetallic compounds with Cd and thus could be separated from the uranium. However, the electrolysis as currently carried out (Figure 2.3) does not employ the liquid Cd as a cathode, but only as an anode to recover any U metal that is scraped off or falls from the steel mandrel and dissolves in the liquid Cd pool.

AN ELECTROMETALLURGICAL APPROACH FOR TREATMENT OF EXCESS WEAPONS PLUTONIUM

Several developments in 1995 caused the DOE to request additional evaluations by the committee. One request was to evaluate the scientific and technological issues associated with extending ANL's electrometallurgical research and development program to handle plutonium, in the event that DOE might pursue an electrometallurgical treatment option for the disposition of excess weapons plutonium (WPu).

Initially, the electrometallurgical process under development at ANL (see Figure 2.2) was designed to separate actinide elements from fission products and direct them into separate output streams. The actinide-fission product separation was to occur during electrorefining as a consequence of the different oxidation-reduction properties of two different cathodes. Relatively pure uranium was to be deposited at a steel cathode, and the transuranic fraction was to be collected at a molten cadmium cathode.

BACKGROUND AND DEVELOPMENT OF ELECTROMETALLURGICAL TECHNOLOGY

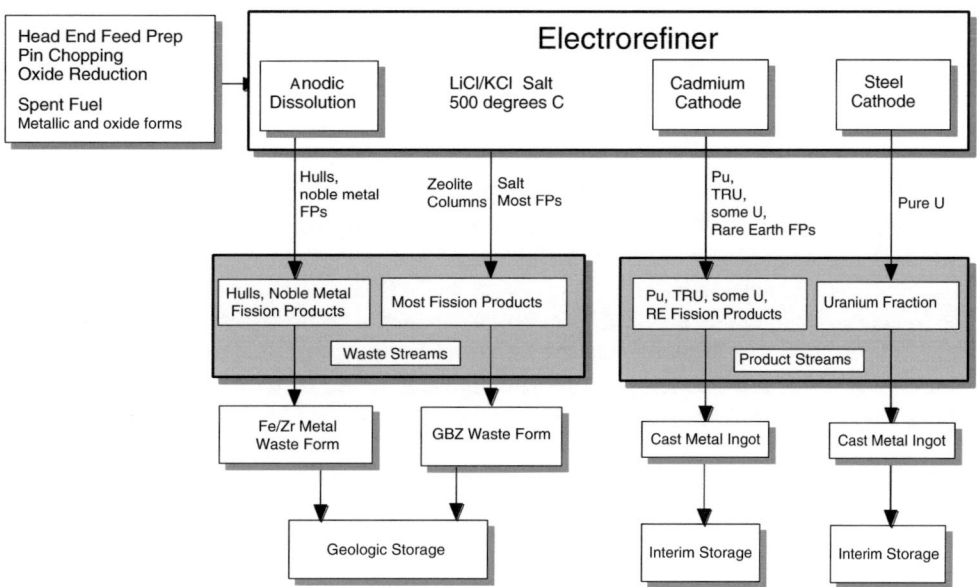

FIGURE 2.2 Original ANL electrometallurgical process scheme. In adaptation of the process for treatment of plutonium, the plutonium would be introduced at the point denoted *spent fuel, metallic and oxide forms.*

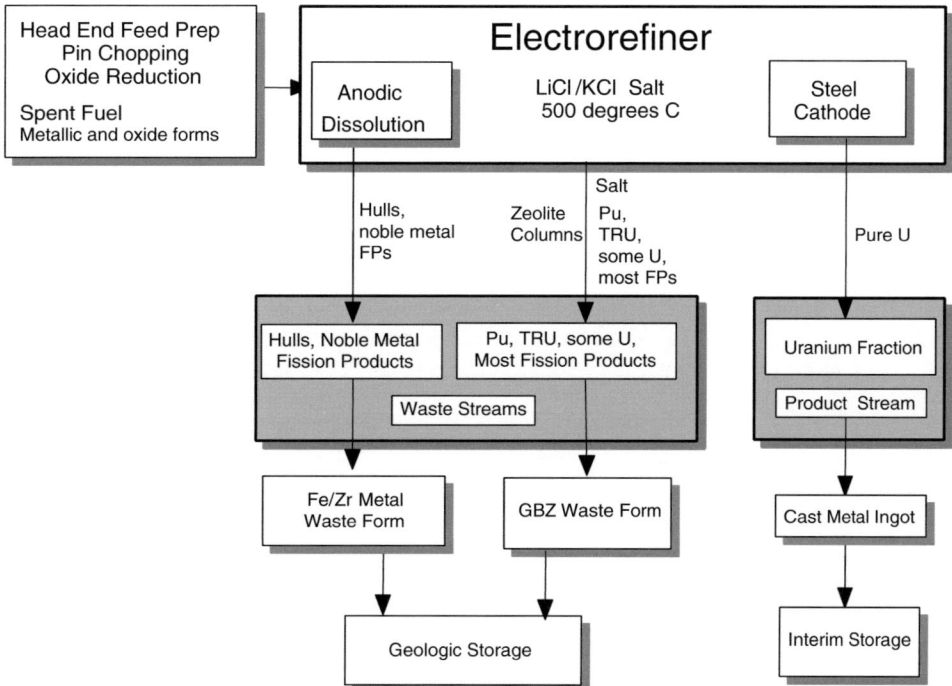

FIGURE 2.3 Current ANL electrometallurgical process scheme, in which the cadmium cathode has been removed. In adaptation of the process for treatment of plutonium, the plutonium would be introduced at the point denoted *spent fuel, metallic and oxide forms.*

In 1995 ANL modified its flow sheet by eliminating the cadmium cathode, with the consequence that the transuranic elements would remain in the molten salt along with the fission products, ultimately to be incorporated into the waste form derived from zeolite. The co-location of plutonium (and the other transuranics) with highly radioactive fission products suggested that the process might also be employed for disposition of excess weapons plutonium (WPu).

Earlier, in 1994 and 1995, the National Academy of Sciences (NAS) Committee on International Security and Arms Control (CISAC)[12] and its associated Panel on Reactor-Related Options for the Disposition of Excess Weapons Plutonium[13] had evaluated options for the disposition of plutonium. CISAC introduced the concept of the "spent fuel standard" to describe a "condition in which the WPu has become roughly as difficult to acquire, process, and use in nuclear weapons as it would be to use plutonium in commercial spent fuel for this purpose."[14] Since the amount of spent fuel was growing rapidly, the panel further concluded that "there would be very little security gain from special efforts to completely eliminate the WPu, or render it much less accessible even than the plutonium in spent fuel, unless society were prepared to take the same approach with the global stock of civilian plutonium."

In its evaluation of EMT (generic pyroprocessing as described above) as an alternative approach for plutonium disposition, the Panel on Reactor-Related Options cited several disadvantages that it felt effectively excluded the electrometallurgical technique as a viable option in the near term. The panel concluded that the "pyroprocessing approach is not competitive with either vitrification in borosilicate glass or the use of mixed uranium-plutonium oxide fuel (MOX) in existing reactors, both of which would be likely to involve lower costs, lower technical uncertainties, and shorter delay."[15]

The DOE sought advice on whether the CISAC conclusions remained valid in view of ANL's subsequent modification of the EMT process (Figure 2.3) to capture the plutonium and other transuranic elements in a zeolite matrix along with most of the fission products. For the possible application of the electrometallurgical treatment technology to surplus fissile material disposition, ANL also proposed the addition of CsCl from capsules at DOE's Hanford Reservation to create the radiation barrier to meet the "spent fuel standard."

In its response to DOE, the Committee on Electrometallurgical Techniques for DOE Spent Fuel Treatment concluded that disposition of WPu would involve different feeds for use in SNF processing, raising several concerns with respect to electrometallurgical processing, zeolite loading, and waste form performance. Although ANL had at that time demonstrated an initial program in evaluating zeolite loading, considerable work would be needed to demonstrate this step in a large-scale, continuous operation with fully radioactive loadings on zeolite columns.

Introduction of WPu in the EMT process would significantly increase the demands on the technology to meet the performance requirement for waste forms relative to the use of the waste forms for ultimate disposal of fission products from SNF processing. These considerations led the committee to recommend that "greater priority should be given to the development of a strategy and a relevant test protocol to demonstrate acceptability of waste forms. This activity is of the highest importance relative to all other aspects in the development of the electrometallurgical technique for WPu disposition" (p. 8).[16]

---

[12]National Academy of Sciences, Committee on International Security and Arms Control (CISAC), *Management and Disposition of Excess Weapons Plutonium,* National Academy Press, Washington, D.C., 1994.

[13]National Academy of Sciences, Panel on Reactor-Related Options for the Disposition of Excess Weapons Plutonium, Committee on International Security and Arms Control (CISAC), *Management and Disposition of Excess Weapons Plutonium: Reactor-Related Options,* National Academy Press, Washington, D.C., 1995.

[14]National Academy of Sciences, Panel on Reactor-Related Options for the Disposition of Excess Weapons Plutonium, Committee on International Security and Arms Control (CISAC), *Management and Disposition of Excess Weapons Plutonium: Reactor-Related Options,* National Academy Press, Washington, D.C., 1995, p. 73.

[15]National Academy of Sciences, Panel on Reactor-Related Options for the Disposition of Excess Weapons Plutonium, Committee on International Security and Arms Control (CISAC), *Management and Disposition of Excess Weapons Plutonium: Reactor-Related Options,* National Academy Press, Washington, D.C., 1995, p. 221.

[16]National Research Council, *An Evaluation of the Electrometallurgical Approach for Treatment of Excess Weapons Plutonium,* National Academy Press, Washington, D.C., 1996.

The committee concurred with the earlier statements of CISAC and its Reactor Panel on excess weapons plutonium: "The existence of this surplus material constitutes a clear and present danger."[17] "The timing of disposition options is crucial to minimizing risks."[18] The urgency of moving ahead with disposing of weapons plutonium made scheduling considerations an important factor in deciding whether or not the electrometallurgical technique would be a practicable and timely solution.

The committee also noted that the potential advantage of the electrometallurgical technique for disposition of excess plutonium would depend on the availability of operational electrometallurgical process equipment. In the absence of such equipment, EMT would not be a viable approach to disposition of plutonium. The committee further concluded that until successful completion of the EBR-II fuel demonstration and treatment of additional spent fuel had been undertaken, it would be imprudent to plan for use of the electrometallurgical technique for disposition of weapons plutonium.

These conclusions led the committee to make two overall recommendations in its third report:[19] First, "a decision on the use of the electrometallurgical technique for weapons plutonium disposition cannot be made until the demonstration of this technology shows whether or not this process is viable for treating DOE spent fuels. If a weapons plutonium disposition technology is to be selected for use with weapons pits before the electrometallurgical technology demonstration program is concluded, this committee recommends that the electrometallurgical technique not be included as a candidate technology" (p. 8), and second, "the potential of the electrometallurgical technique as an adjunct for long-term disposition of non-pit excess plutonium remains a possibility, but the technology is still at too early a stage of development to be evaluated relative to disposition alternatives such as glass or MOX [mixed uranium-plutonium oxide fuel]" (p. 8). Any decision to use electrometallurgical technology for WPu disposition remains dependent on establishing acceptance criteria for the waste forms.

---

[17]National Academy of Sciences Committee on International Security and Arms Control (CISAC), *Management and Disposition of Excess Weapons Plutonium,* National Academy Press, Washington, D.C., 1994, p. 1.

[18]National Academy of Sciences, Panel on Reactor-Related Options for the Disposition of Excess Weapons Plutonium, Committee on International Security and Arms Control (CISAC), *Management and Disposition of Excess Weapons Plutonium: Reactor-Related Options,* National Academy Press, Washington, D.C., 1995, p. 2.

[19]National Research Council, *An Evaluation of the Electrometallurgical Approach for Treatment of Excess Weapons Plutonium,* National Academy Press, Washington, D.C., 1996.

3

The Electrometallurgical Process at Argonne National Laboratory

INTRODUCTION

The electrometallurgical process, as utilized in ANL's demonstration project, consisted of a series of distinct steps to process SNF. As the demonstration project progressed, the EMT process was streamlined to generate three product streams. The details of the equipment used in the EMT process, and the waste forms and products that result from this process, are the focus of this chapter.

Figure 3.1 shows the electrometallurgical processing equipment associated with each of the EMT unit operations. The electrometallurgical treatment process for driver and blanket fuel comprises four stages: disassembly of the fuel elements with an element chopper; electrochemical removal of the uranium from the stainless-steel-clad fuel element pieces in an electrorefiner; consolidation of the electrodeposited uranium and removal of the entrained salt in a cathode processor; and finally, casting of the uranium product from the cathode processor into an ingot in a casting furnace. Down blending of the highly enriched uranium (HEU) driver fuel with depleted uranium takes place in both the cathode processor and the casting furnace. For blanket fuel, which is mostly depleted uranium, down blending is not necessary.

Two different electrorefiners are used to treat the segmented fuel elements. The Mark-IV electrorefiner employs a relatively simple mechanical design and has sufficient throughput to treat the relatively small amount of sodium-bonded driver fuel in the ANL-W inventory (3.5 metric tons of heavy metal), whereas the more sophisticated Mark-V electrorefiner with a much higher throughput is designed to treat the more plentiful sodium-bonded blanket fuel (56.5 metric tons). A detailed description of these electrorefiners as well as the equipment and procedures employed at each stage of the EMT treatment process is given below.

Fuel Element Choppers

The driver fuel assembly (Figure 3.2) consists of a hexagonal stainless-steel tube (cladding) containing 61 fuel elements loaded with HEU fuel.[1] This fuel is a uranium alloy with 10 wt % Zr. The uranium consists of 63% atomic ^{235}U after burnup. The gaps between the fuel and the inside walls of the cladding are filled with sodium to

---

[1]The assemblies used during the demonstration project contained 61 fuel elements. There are additional fuel assemblies stored at ANL-W that contain 91 fuel elements.

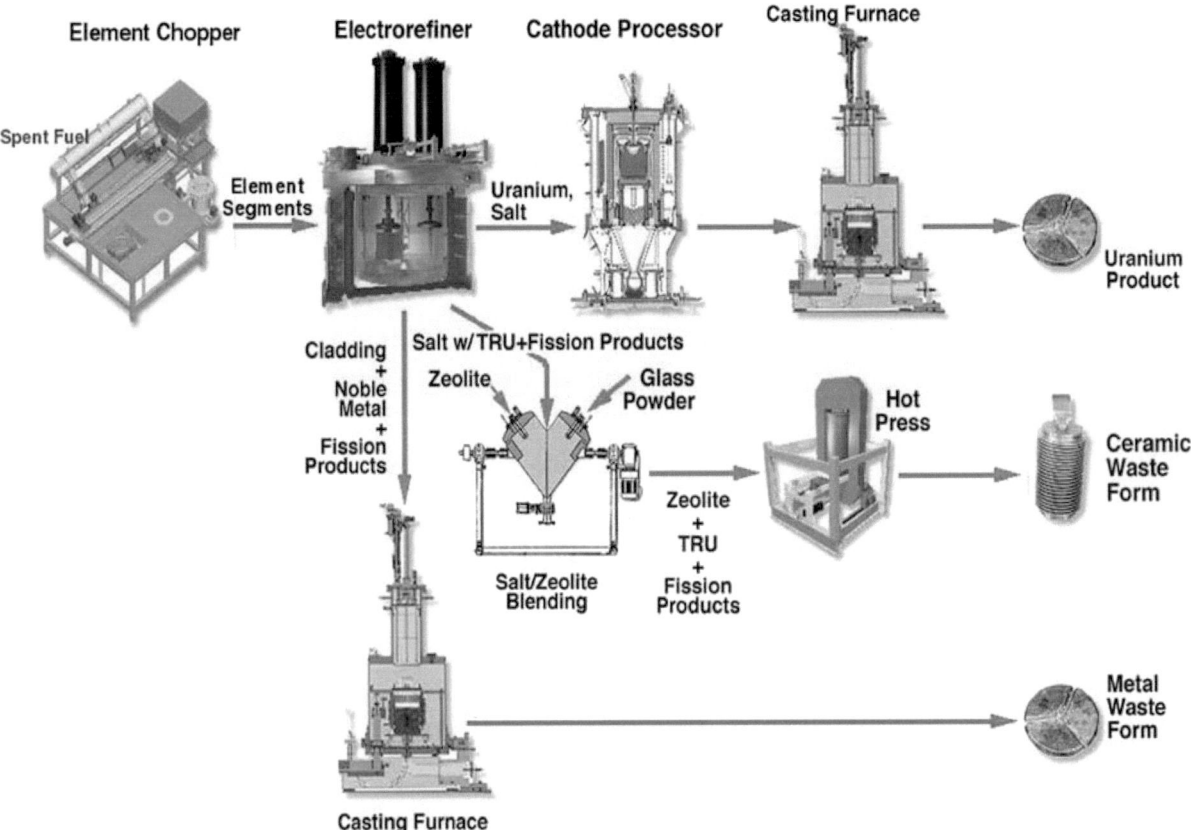

FIGURE 3.1 Equipment used in the treatment of EBR-II spent nuclear fuel.
SOURCE: Argonne National Laboratory.

provide a thermal bond; sufficient sodium-metal is also added to cover the top of the fuel. An argon-gas-filled expansion region extends from the top of the fuel/sodium to the top of each fuel element. An EBR-II driver assembly typically contains 4.5 kg of uranium.

The EBR-II blanket fuel is depleted uranium in a stainless-steel cladding (Figure 3.3). Each assembly contains 19 fuel elements; each element contains 5 slugs of uranium, each weighing 0.50 kg, for a total of 2.5 kg of uranium per element.[2] The total weight of uranium in a blanket assembly is normally 47.5 kg.[3] As is the case for the driver fuel, the gap between the fuel and the cladding is filled with sodium. The average discharge burnup for this fuel was approximately 1.2 wt %.

---

[2]K.M. Goff, L.L. Briggs, R.W. Benedict, J.R. Liaw, M.F. Simpson, E.E. Feldman, R.A. Uras, H.E. Bliss, A.M. Yacout, D.D. Keiser, K.C. Marsden, and C.W. Nielsen, *Production Operations for the Electrometallurgical Treatment of Sodium-Bonded Spent Nuclear Fuel*, NT Technical Memorandum No. 107, Argonne National Laboratory, Argonne, IL, 1999, p. 13.

[3]S.R. Sherman, D. Vaden, R.D. Mariani, B.R. Westphal, T.S. Bakes, S.S. Cunningham, B.A. Johnson, D.V. Laug, and J.R. Krsul, *Process Description for Blanket Fuel Treatment Operations*, NT Technical Memorandum No. 113, Argonne National Laboratory, Argonne, IL, 1999, p. 36.

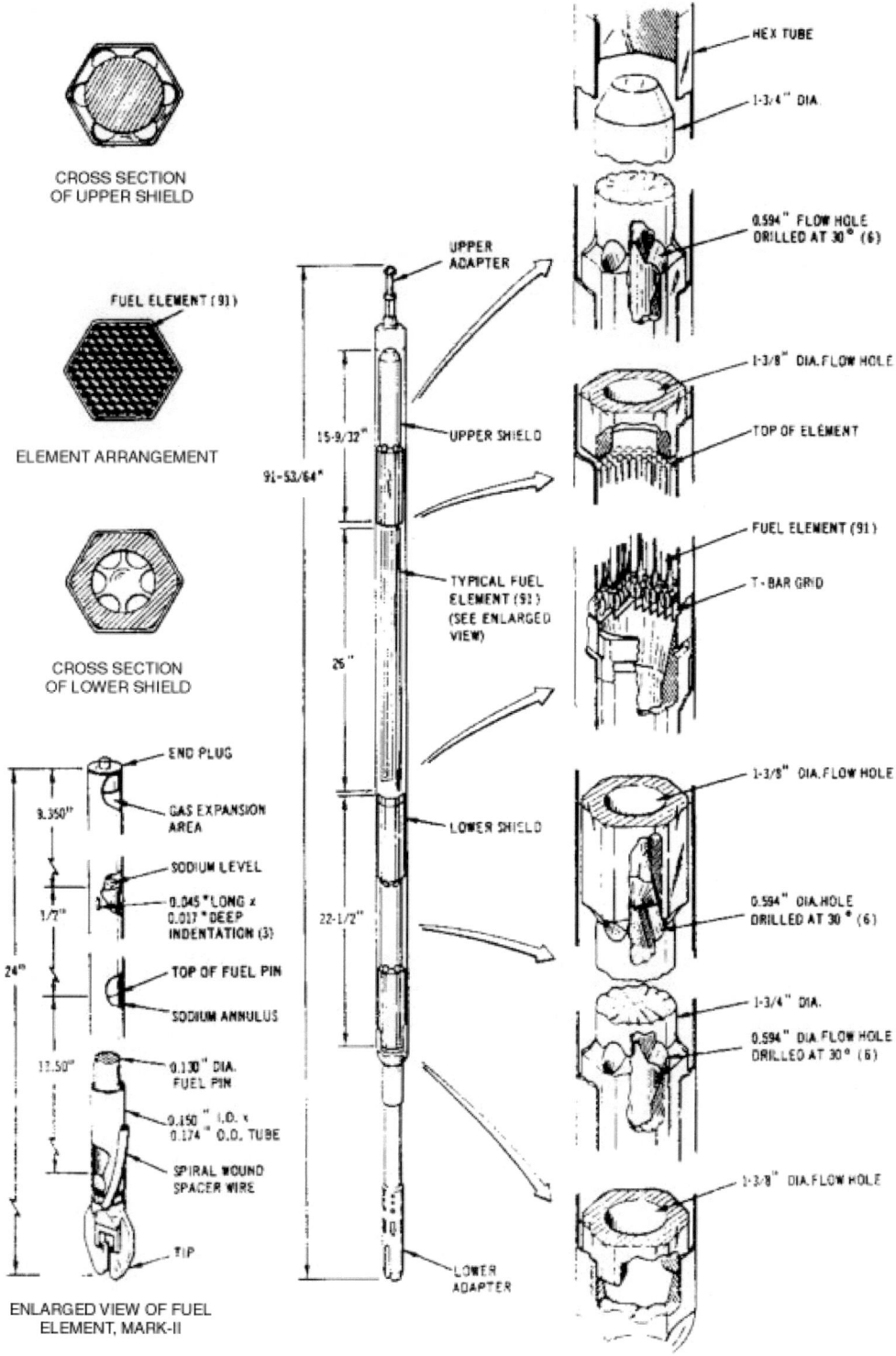

FIGURE 3.2 Driver fuel assembly.
SOURCE: Argonne National Laboratory.

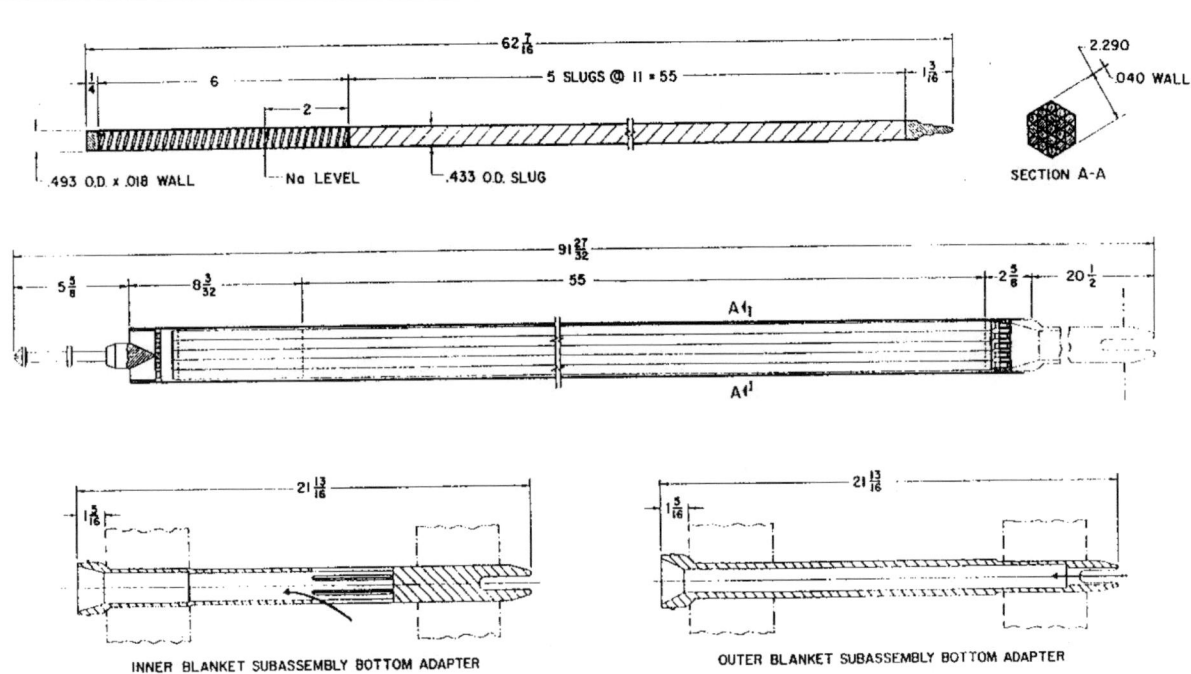

FIGURE 3.3 EBR-II inner and outer blanket fuel assemblies.
SOURCE: Argonne National Laboratory.

The fuel element choppers are pneumatic punch presses that have been modified with blades for shearing driver and blanket fuel elements into segments for loading into the anode compartments of the Mark-IV and Mark-V electrorefiners. The element choppers used for driver and blanket fuel differ somewhat in design but perform essentially the same function. The driver element chopper is programmed to chop that portion of the fuel element containing the fuel and sodium bond into pieces for loading into the fuel dissolution basket of the Mark-IV electrorefiner. The chopped blanket fuel segments are fed directly into the anode basket assemblies of the Mark-V anode-cathode module. As part of ANL's demonstration project criterion that required a blanket throughput rate of 150 kg per month sustained for one month,[4] the blanket element chopper was used to process 3.5 blanket fuel assemblies or 66 blanket fuel elements for a total of 164.4 kg of uranium.

Mark-IV Electrorefiner

The Mark-IV electrorefiner (ER; Figure 3.4) is designed specifically for treating the EBR-II driver fuel. It consists of a 1.0 m diameter, 1.0 m deep 2.25 Cr-1 Mo steel vessel with a cover. Four ports (approximately 25.4 cm. diameter) exist in the top of the ER for anode and cathode assemblies. The vessel is normally filled to a depth of 33 to 35.6 cm with molten LiCl-KCl eutectic and to about 10 cm with molten cadmium metal, which forms a pool in the bottom of the electrorefiner vessel. The cell is operated at 500 °C. There are four 25.4-cm (10-in.) ports

---

[4]The full success criteria for the demonstration project, along with the goals to meet them, are included in Chapter 6.

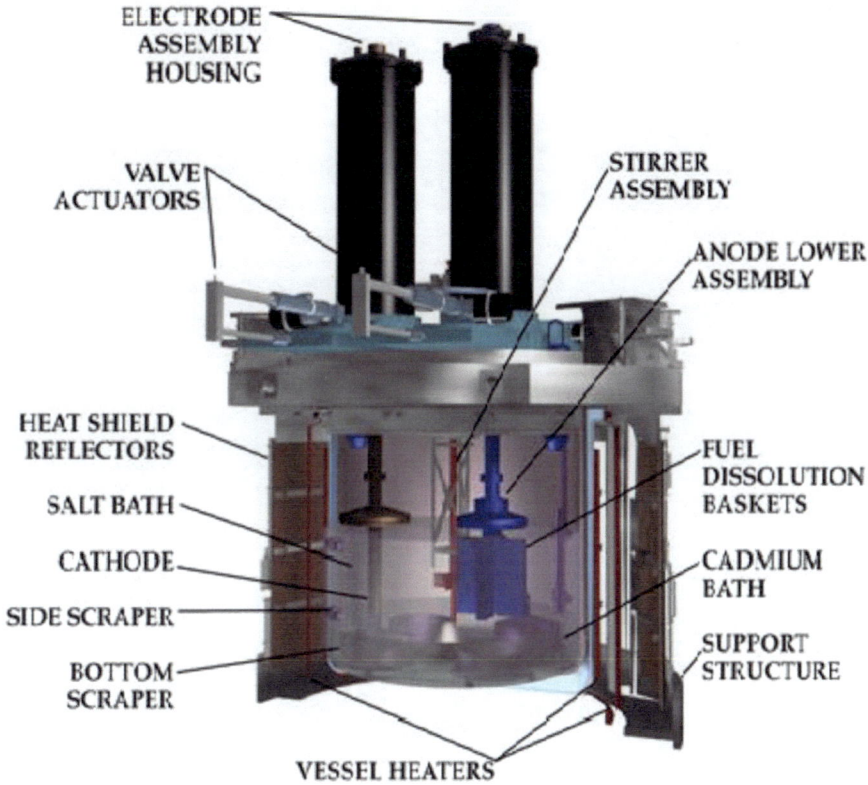

FIGURE 3.4 Schematic of the Mark-IV electrorefiner.
SOURCE: Argonne National Laboratory.

in the Mark-IV ER cover to permit electrode assemblies to be inserted in the molten salt electrolyte. Two ports are used for anodes and two are used for cathodes.

The anode assembly used in the Mark-IV ER consists of four rectangular, perforated stainless-steel compartments, called fuel dissolution baskets, arranged in the shape of a cross (Figure 3.5). Each anode assembly (four baskets) can hold about 8 kg of uranium, which is roughly the amount of uranium found in two driver fuel assemblies. The cathode is simply a mild steel rod or mandrel. The stainless-steel mandrel is normally immersed in the molten salt to a depth of 23 cm during the electrorefining of uranium. Both the anode assembly and the cathode mandrel are rotated to provide convective mass transport when the cell is in operation.

Steel scrapers are mounted on the inside wall of the electrorefiner to control the growth of the dendritic electrodeposit. These scrapers are placed near each cathode opening to prevent the uranium deposit from exceeding 25.4 cm in diameter; otherwise, it would be impossible to remove the uranium deposit through the cathode port at the conclusion of an electrorefiner run. Bottom scrapers are also necessary; they prevent the dendritic uranium deposit from contacting the cadmium pool in the bottom of the electrorefiner. During operation, the scrapers dislodge some of the uranium deposit, which falls into the cadmium pool and dissolves. A small stirring propeller inserted through the ER cell cover into the cadmium pool provides agitation.

The Mark-IV ER's overall anode batch size of 16 kg, achieved by using dual anode assemblies in parallel with a single serial cathode. The efficiency of the overall electrorefining operation is enhanced by using a second

FIGURE 3.5 Fuel dissolution baskets.
SOURCE: Argonne National Laboratory.

cathode inserted into the molten salt through the fourth port. This second cathode allows the product from a direct transport run to be harvested from the off-line mandrel while the other serves as the cathode during a second transport run.

Cadmium chloride is added to the molten salt in the ER at the beginning of each run to oxidize some of the uranium metal in the chopped fuel segments to U^{3+}; U^{3+} must be present in the melt to promote initial deposition at the cathode. Additions of $CdCl_2$ are needed periodically to maintain the U^{3+} concentration at or above this level. The cell is also fitted with a continuously operated cadmium vapor trap to reduce the accumulation of cadmium on the electrode components.

The Mark-IV electrorefiner can be operated in three modes: (1) direct transport, (2) anodic dissolution, and (3) deposition. In the direct transport, the uranium is electrochemically oxidized from the chopped fuel elements as U^{3+}, transported through the molten salt bath by forced convection, and then reduced to the metal at the stainless-steel mandrel. In the anodic dissolution mode, the fuel dissolution baskets serve as the anode, and the cadmium pool serves as the cathode: the uranium is electro-chemically oxidized from the chopped fuel elements as U^{3+}, transported through the melt by forced convection, and then reduced at the surface of the cadmium pool, where it dissolves in the cadmium. (Reduction of plutonium at the cadmium cathode did not occur because of the low concentration of plutonium in the salt and careful voltage regulation.) In the deposition mode, the cadmium pool serves as the anode, and the stainless-steel mandrel serves as the cathode. The uranium is oxidized from the cadmium pool and deposited on the steel mandrel. Anodic dissolution and deposition may thus be regarded as a two-step process. In the deposition mode, any uranium dendrites or pieces that dislodge from the mandrel dissolve in the cadmium and are oxidized and redeposited on the mandrel. The direct transport and the deposition mode were both used in ANL's electrometallurgical demonstration project.

As part of ANL's demonstration project,[5] the Mark-IV electrorefiner was used to treat twelve driver assemblies over a two-month period at an average rate of 24 kg of uranium per month compared to the target criterion of 16 kg (~4 driver assemblies) per month over a three-month period.[6]

Mark-V Electrorefiner

The processing capacity of the Mark-IV ER is inadequate to permit timely processing of blanket fuel, which exists in much larger quantities than driver fuel. Thus, the low processing capacity of the Mark-IV ER was the driving force for the development of the Mark-V system, which was designed for processing of EBR-II blanket fuel. The basic difference between these two electrorefiners is the design of the anodes and cathodes, which allowed significant increase and throughput.

The 25.4-cm (10-in.) ACMs used in the Mark-V ER are derived from the 20.3-cm (8-in.) prototype ACM that was first tested in the Mark-III ER at ANL-E and a much larger 63.5-cm (25-in.) prototype ACM under exploration at ANL-E for use in a high-throughput electrorefiner (HTER)[7] with an anode batch size of 150 kg. However, the HTER has been beset with a number of technical problems such as cell shorting and rotor stalling due to the buildup of a dense uranium deposit between the anode and cathode surfaces of the HTER, and it requires additional development (see Chapter 5). ANL apparently intends to use an ER based on this HTER to process fuel other than that associated with the EBR-II reactor.

Figure 3.6 is schematic of the Mark-V ER in use in the Fuel Conditioning Facility (FCF) at ANL-W. The Mark-V electrorefiner vessel is identical in dimensions to the Mark-IV vessel and holds approximately 650 kg of LiCl-KCl eutectic. Like the Mark-IV system, the Mark-V ER vessel has a cover equipped with four 25.4-cm-diameter ports for the insertion of electrodes. However, the Mark-V ER contains four integrated, independently operated anode-cathode modules (ACMs) that are considerably more complex than the simple rotating anode baskets and stainless-steel mandrel cathodes used in the Mark-IV system. Each ACM consists of two rotating anode basket assemblies with a total of nine anode baskets and three stationary cathode tubes (outer, middle, and inner) arranged in a concentric, circular configuration (Figure 3.7). Attached to the bottom of each ACM is a product collector basket. As the uranium deposit accumulates on the cathode surfaces, a series of closely spaced beryllia scraper blades mounted on the leading edges of the anode baskets remove the uranium deposits, which fall into the removable screen buckets positioned under the ACMs. The optimum position of these scrapers (leading or trailing), the best spacing between the scrapers and the cathode surfaces, and the best material to use for the fabrication of these scrapers have been determined by extensive trial-and-error experimentation. Each ACM has a capacity of about 37 kg of uranium, and each has its own 600-ampere power supply. With ACMs installed in all four ports of the Mark-V ER, the total anode batch size approaches 150 kg. This batch size is considerably greater than the 16-kg batch size of the Mark-IV ER and is roughly equivalent to about three blanket fuel assemblies. However, only three ACMs were installed in the ER to meet the ANL-W demonstration goal of 150 kg uranium/month for blanket fuel processing.

Before operation of the Mark-V ER can begin, the ACMs must be charged with fuel segments produced by the blanket element chopper. After the ACMs are placed into the electrorefiner, the anode baskets are rotated counterclockwise at 10 to 20 rpm while the cathode tube assembly and product collector are held stationary. For the

---

[5]The full success criteria for the demonstration project, along with the goals to meet them, are included in Chapter 6.
[6]R.D. Mariani, D. Vaden, B.R. Westphal, D.V. Laug, S.S. Cunningham, S.X. Li, T.A. Johnson, J.R. Krsul, and M.J. Lambregts, *Process Description for Driver Fuel Treatment Operations*, NT Technical Memorandum No. 111, Argonne National Laboratory, Argonne, IL, 1999.
[7]E.C. Gay, S.R. Sherman, J.L. Willit, and R.K. Ahluwalia, *Development of the Electrorefining Process for Blanket Fuel*, NT Technical Memorandum No. 114, Argonne National Laboratory, Argonne, IL, 1999, p. 10.

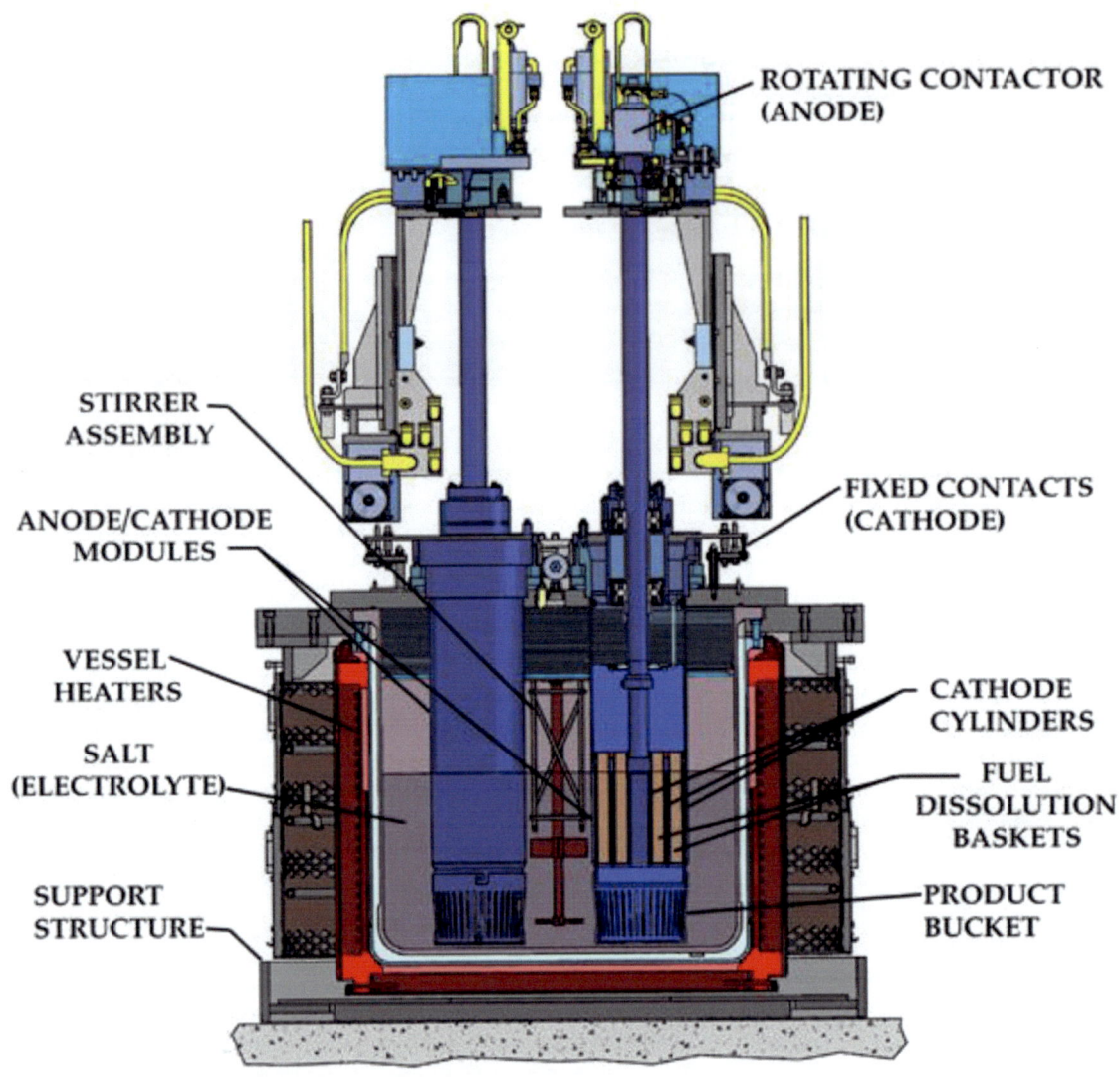

FIGURE 3.6 Schematic of the Mark-V electrorefiner.
SOURCE: Argonne National Laboratory.

demonstration project, UCl_3 was added to the melt at a concentration of 4 to 7 wt %. Adding UCl_3 and maintaining a steady-state concentration of it serve the same purpose as that described above for the Mark-IV ER. Because the Mark-V ER does not utilize a cadmium pool, it operates only in the direct transport mode. Thus, electrorefining is initiated by imposing a current between the anode and cathode surfaces of an ACM; this results in the direct transport of uranium from the fuel segments in the anode basket assembly to the cathode surfaces of the ACM.

For smooth ACM operation, it is optimal if the uranium forms on each cathode surface as loosely adherent dendrites so that it can be removed easily by the beryllia scrapers. Unfortunately, in addition to dendrites, dense deposits of uranium invariably occur that cannot be removed by the scrapers. If allowed to develop beyond a

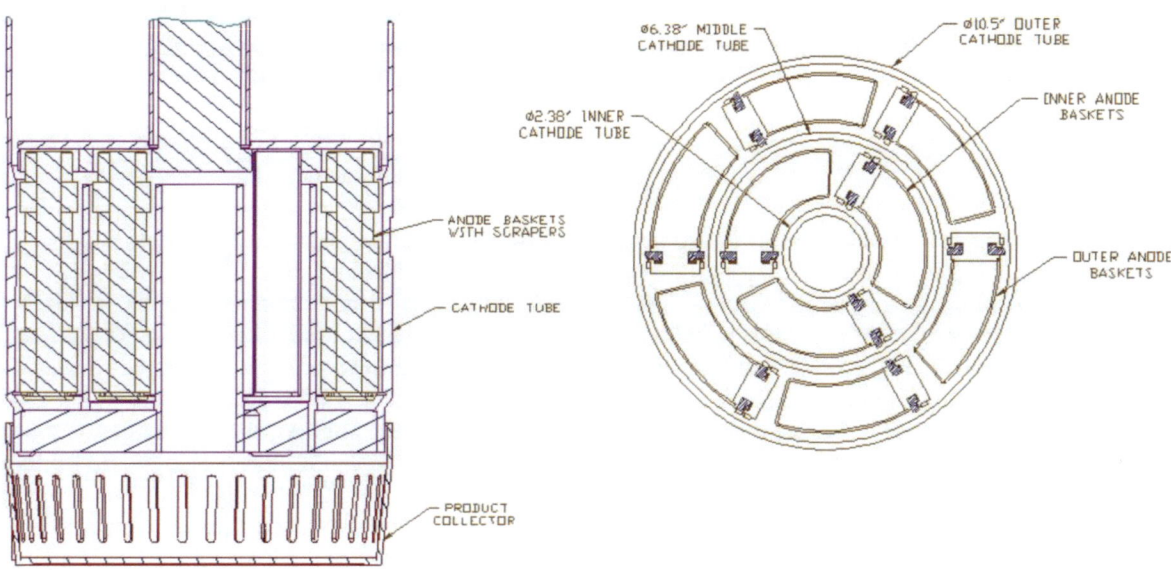

FIGURE 3.7 Anode-cathode configuration.
SOURCE: Argonne National Laboratory.

certain thickness, these dense uranium deposits can stall ACM rotation. It was determined that such deposits must be removed after about 200 ampere-hours of direct transport deposition to ensure continued successful operation of the Mark-V ER. These deposits are removed by electrochemical stripping, i.e., by reversing the direction of the current so that the cathode tubes become anodes and vice versa. A washing step is imposed between each deposition and stripping step. Washing consists of rotating the anode basket assembly in the ACM without the passage of current. An automated sequence involving a stripping step, washing step, deposition step, and a second washing step constituting an operational cycle appears to have mitigated the problem,[8] with some decrease in overall efficiency.

The material that collects in the product collector basket consists of uranium and approximately 20 wt % salt. When the collector basket is removed from the electrorefiner, the uranium-salt mixture forms a dense mass that is tightly bound to the collector basket and cannot be removed by simply inverting the basket. Two methods were developed to remove the adherent product mass from the collector basket. The first utilizes a cutting/grinding tool, called the product-harvesting tool, to dig the solidified product out of the product collector basket. The second involves placing the product collector basket in a small oven, allowing the molten salt to drain away, and then inverting the product collector basket over an intermediate container. The material harvested from the product collector basket is sent to the cathode processor for further treatment.

Each ACM in the Mark-V ER is capable of processing about 87 to 100 kg of uranium per month. The use of additional ACMs does not necessarily lead to a proportional increase in the monthly process rate of uranium, because only one ACM can be serviced at a time. Therefore, anode basket and product collector changeover must be carefully coordinated with the start of each run to avoid conflicts that might arise from the need to service two ACMs simultaneously. As part of ANL's demonstration project,[9] the Mark-V ER processed the equivalent of 4.3 blanket assemblies (206 kg U per month), corresponding to a total uranium deposit mass of 204.9 kg. During this

---

[8]This sequence is discussed in detail in Appendix B.
[9]The full success criteria for the demonstration project, along with the goals to meet them, are included in Chapter 6.

time, 1.37 ACMs were active and the product collector rate was 212 g uranium/h per ACM. The average product collector fill rate per ACM limits the total processing rate. Each product collector can hold 13 kg of uranium product (uranium + 20 wt % salt), and three product collectors are required during the electrorefining of every 10 blanket elements.

Cathode Processor

The purpose of the cathode processor (Figure 3.8) is twofold: to remove entrained salt (and any cadmium) from the uranium electrodeposits by evaporation and to consolidate dendritic deposits. In the case of the driver fuel, depleted uranium must be added to the cathode processor to reduce the enrichment of the ER product to less than 50% for security reasons; this down blending is not necessary for blanket fuel (see Chapter 4). The cathode processor consists of an outer vessel; an induction-heated furnace assembly with coils, liner, and insulation; a crucible assembly with a graphite process crucible and cover, a radiation shield, a condenser, and a receiver crucible. A graphite furnace liner acts as the susceptor in the induction furnace, and the induction coils are passively cooled and protected by a vapor barrier. Entrained salt and any cadmium present in the driver fuel are removed from the molten uranium ER product by vacuum distillation and then deposited as a liquid in the receiver crucible. For convenience, the cathode processor is elevated above floor level so that the crucible assembly can be bottom loaded into the induction furnace. This position permits the process crucible to be loaded, emptied, and cleaned without affecting the furnace assembly.

The graphite process crucible used in the cathode processor is coated with a protective coating of ZrO_2. Between 50 and 100 g of ZrO_2 are applied, and the crucible is baked at 600 °C. The application of excess ZrO_2 leads to the formation of dross and the subsequent loss of the uranium product due to the reaction

$$ZrO_2 + U \rightarrow UO_2 + Zr.$$

The crucible must be cleaned and coated after each run. The use of a beryllia crucible is still under investigation to reduce the formation of dross.

The cathode processor sequence involves loading the crucible with product from the ER, positioning the crucible on the crucible assembly, and transporting the entire assembly into the cathode processor vessel. The cathode processor vessel is evacuated and heated in stages to a maximum temperature of ~1200 °C over a period of approximately 10 hours. ANL determined that operation of the cathode processor at temperatures above 1200 °C drastically increases dross formation and that holding the process crucible temperature slightly below the melting point of the uranium (mp = 1135 °C) for a short time improves the distillation of salt (and cadmium, if present). Thus, the crucible temperature is held at 1100 °C for one hour under vacuum before being increased to 1200 °C for one-half hour. After the process crucible has been held at 1200 °C for one-half hour, the current to the induction coils is discontinued, and the cathode processor is allowed to cool under vacuum to ambient temperature.

The final stage of the cathode processor operation involves lowering the process crucible assembly out of the furnace, placing it in a dumping fixture, and dislodging the resulting uranium ingot from the crucible. After several runs, the distillate containing salt and cadmium (Mark-IV) in the receiver crucible is carefully returned to the ER from which it originated so that any cadmium present in the distillate is not inadvertently added to the ER used to treat blanket fuel (Mark-V).

Casting Furnace

The casting furnace (Figure 3.9), which has evolved from earlier induction furnaces designed at ANL-E, provides a means to reduce the ^{235}U enrichment of the driver fuel product from the cathode processor by the addition of depleted uranium resulting from the cathode processor and to consolidate further the uranium product. Its components are similar to those of the cathode processor except that there is no condenser stage and associated receiver crucible to collect the distillate. Like the cathode processor, the casting furnace is based on an induction furnace and a graphite crucible, and it can be evacuated. The furnace has two gas-tight flanges (top and bottom)

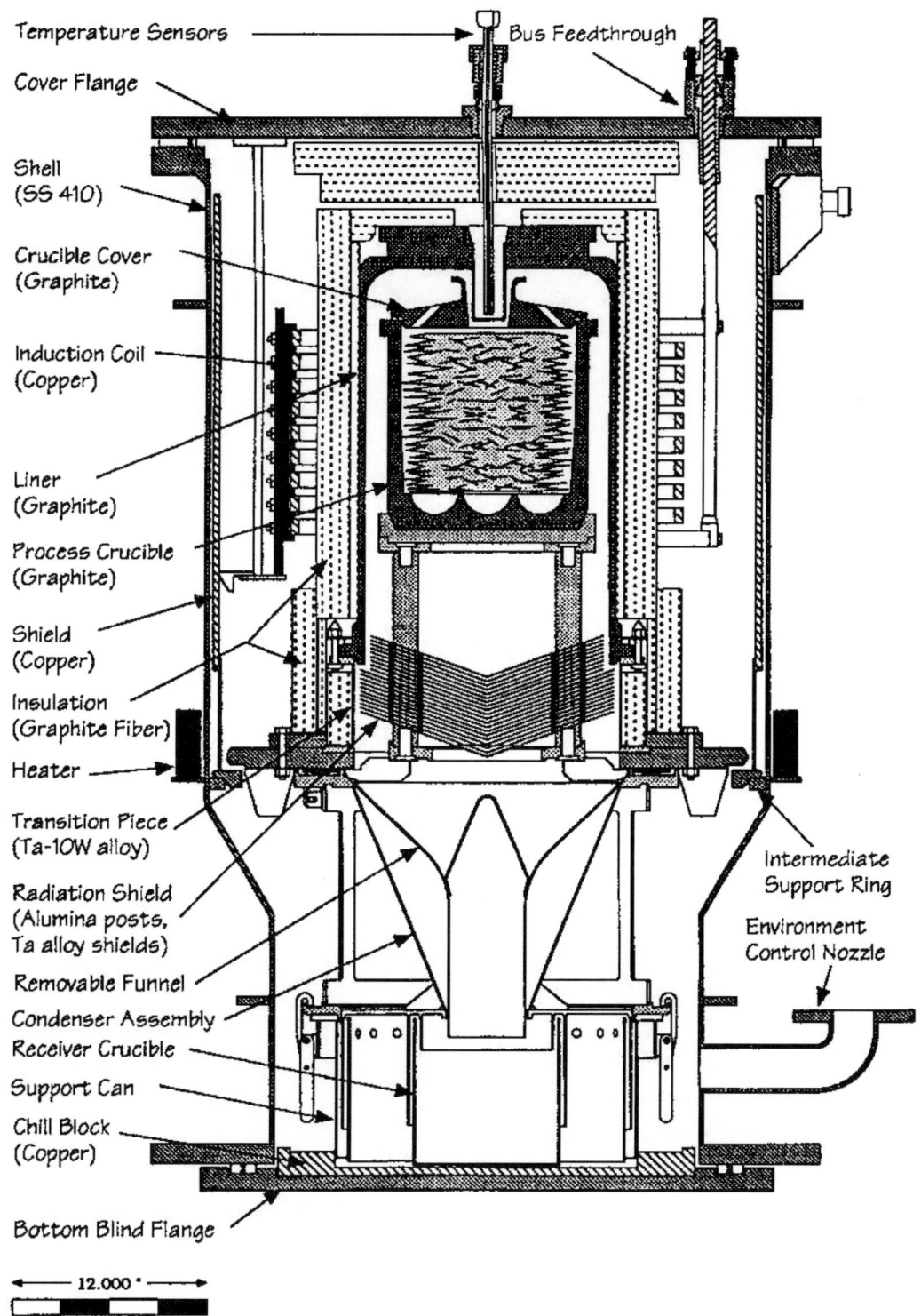

FIGURE 3.8 Schematic of the cathode processor.
SOURCE: Argonne National Laboratory.

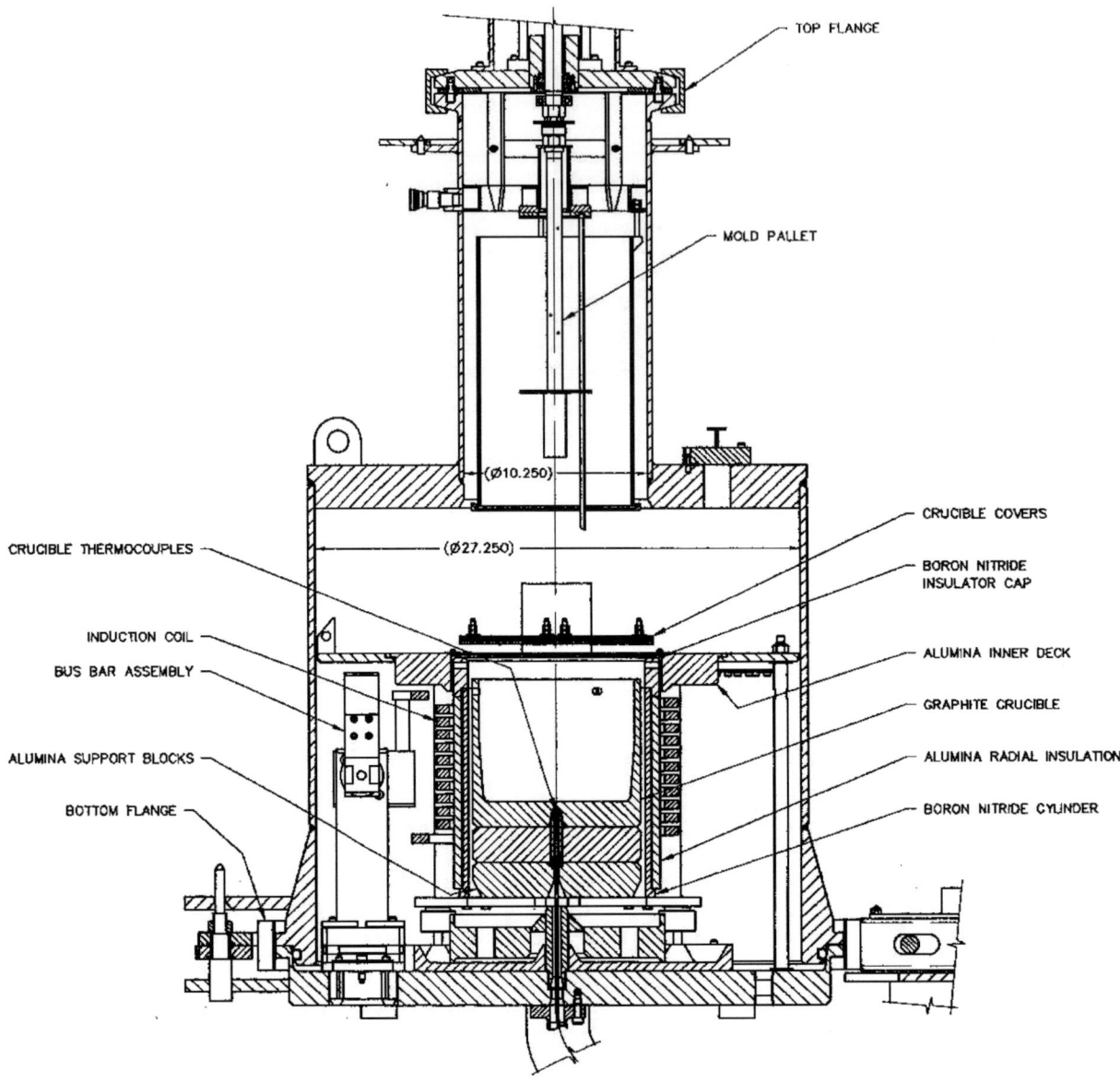

FIGURE 3.9 Schematic of the casting furnace.
SOURCE: Argonne National Laboratory.

that permit access to the casting crucible and other internal components. The crucible is normally loaded through the top flange employing the same fixture used to unload the cathode processor.

The operating parameters associated with the casting furnace include the crucible coating, temperature control, and pressure control. A crucible coating is required in the casting furnace to minimize the interaction of molten uranium with the graphite crucible and to prevent the cast ingot from adhering to the crucible walls. Yttria (Y_2O_3) was found to be suitable for this purpose and is applied in aerosol form. The loss of uranium due to reaction with the Y_2O_3 has been found by ANL to be much smaller than the loss due to reaction with the ZrO_2 crucible lining in the cathode processor crucible.

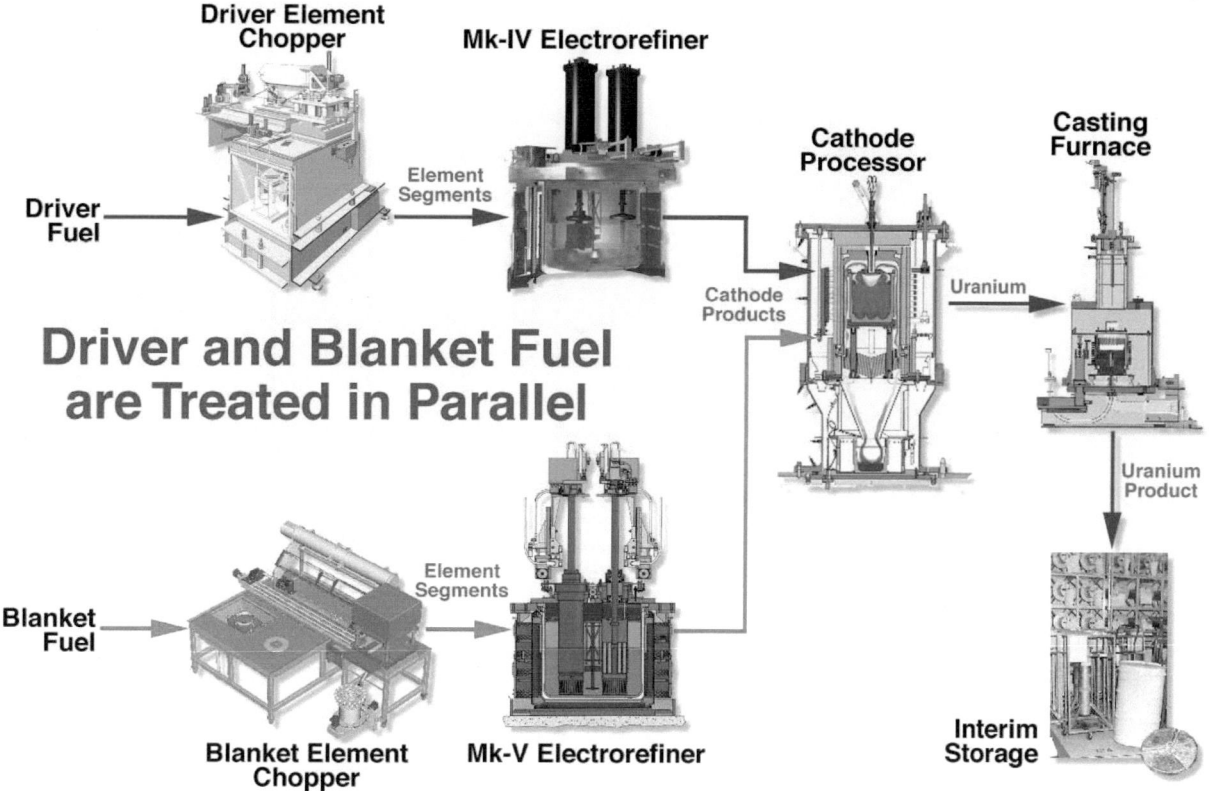

FIGURE 3.10 Flow charts for the processing of driver and blanket fuel.
SOURCE: Argonne National Laboratory.

Figure 3.10 shows flow charts for the processing of blanket and driver fuel during the demonstration phase. The amounts of material that were treated as of July 1999 in the cathode processor and casting furnace are summarized in Appendix B.[10] For the driver fuel, 468 kg of dendritic product (which consist of uranium and approximately 18% of the salt) was produced by the Mark-IV ER. This is the amount of uranium contained in roughly 100 driver assemblies.[11] Likewise, the treatment of 5 of 25 blanket assemblies in the Mark-V ER produced 162.7 kg of electrorefined uranium (including approximately 18% salt).

FABRICATION OF WASTE FORMS

Metal Waste Form

Following the electrorefining operations the stainless-steel cladding hulls are left in the anode basket, along with the noble metal fission products (Zr, Mo, Ru, Rh, Pd, etc.), actinides and adhering salt electrolyte. The uranium content is about 4 wt %.

---

[10]Presentation by Brian R. Westphal to the committee, ANL-W, July 21-22, 1999.
[11]Presentation by Robert W. Benedict to the committee, ANL-W, July 21-22, 1999.

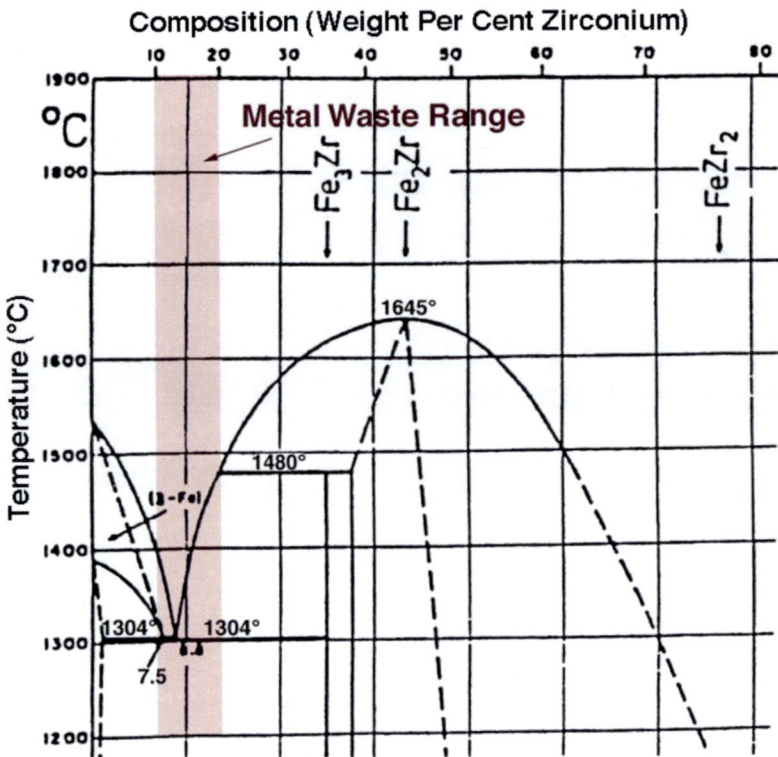

FIGURE 3.11 Iron-zirconium phase diagram.
SOURCE: Argonne National Laboratory.

Zr metal is added to improve performance properties of the final metal waste form and to produce a lower-melting point alloy. The target composition is stainless steel and 15 wt % Zr, with the allowable Zr concentration ranging from 5 to 20 wt %. As shown in the Fe-Zr phase diagram in Figure 3.11, an alloy of 13 wt % Zr is the low melting eutectic.

To distill the adhering processing salts, the material in the anode basket is placed in the cathode processor and heated to 1100 °C. The charge from the cathode processor is placed in an yttrium oxide crucible, is melted at approximately 1600 °C in the casting furnace in an Ar atmosphere, and then is cooled in the crucible or, on the research scale, cast into ingots. The actinides in the metal waste are primarily in this intermetallic phase. The ingot constitutes the metal waste form (MWF).

Ceramic Waste Form

Significant amounts of fission products and transuranic elements accumulate in the LiCl-KCl electrolyte used in the electrorefining process. Although the electrolyte can be recycled, its radioactive components must ultimately be disposed of as high-level radioactive waste (HLW). Vitrification methods used to immobilize other high-level radioactive waste materials cannot be used because glass cannot incorporate high concentrations of salt. For example, borosilicate glass waste forms can accommodate up to 20 mass % Na_2O, but only about 1 mass % Cl. The ceramic waste form (CWF) has been developed to immobilize the active fission products (alkalis, alkaline earths, and rare earths) and transuranic elements of the electrolyte.

The ceramic waste form is produced in a batch process by mixing and blending the waste salt, periodically removed from the electrorefiner, with zeolite 4A at 500 °C to occlude the waste-loaded salt within the cages of the zeolite crystal lattice. The product of this step is called salt-loaded zeolite. Originally, the waste form was made from pure zeolite 4A, which is a fine (~5-mm average particle diameter) powder. Specimens made from this material are called "baseline" glass-bonded sodalite waste forms. In October 1997, a zeolite 4A material with particles nominally ranging from 74 to 250 μm in diameter (as defined by a specified sieve cut of –60+200 mesh) was selected for use in the CWF. It is supplied in a granular form and contains about ten mass percent of a proprietary clay binder. The granular zeolite 4A is much more easily handled; the waste forms made from this material are designated as "reference" glass-bonded sodalite. Salt-loaded zeolite is mixed with a borosilicate glass and consolidated at high temperature (850 to 900 °C) and pressure (14,500 to 25,000 psi) in a hot isostatic press (HIP) to make the final waste form. Under HIP conditions, the salt-loaded zeolite is converted to sodalite. The initial salt-loaded zeolite contained about 12 wt % occluded LiCl/KCl eutectic salt. The amount of occluded salt retained after the transformation in the sodalite phase is assumed to be the same as in the zeolite, an assumption that requires further experimental verification. The standard or reference ceramic waste form is made with 75 mass % salt-loaded zeolite and 25 mass % of a commercial glass that acts as a binder.

The fission product chlorides are allowed to build up in the salt until a sodium chloride limit is reached. The buildup of sodium chloride in the salt raises the melting point of the mixture. The sodium limit was established to provide a safe margin between the liquidus temperature of the salt and the operating temperature. For operation at 500 °C, the limit is estimated to be approximately 6 wt % sodium. Once the sodium limit is reached, enough salt is removed periodically from the electrorefiner to maintain the sodium concentration below the limit. This salt is disposed in the ceramic waste. During the demonstration phase the NaCl limit was not reached.

The steps for producing the ceramic waste form, as developed during ANL's demonstration project, are depicted in Figure 3.12. The equipment shown in this schematic was used for operations with radioactive materials in the Hot Fuel Examination Facility (HFEF) at ANL-W. The demonstration-scale ceramic waste equipment was first tested out of cell with nonradioactive surrogates to develop operating parameters, because relatively little time was available for tests with radioactive materials before completing the demonstration.

A source of dried zeolite was needed for the ceramic waste form. The zeolite used is commercially available from Union Carbide Corporation/UOP, but it contains as much as 22 wt % moisture. A drying cycle was developed by ANL for 40 kg batches of zeolite. Thirty batches of zeolite were dried to less than 1 wt % using a 12-hour heating cycle.

The salt disposed of in the ceramic waste must be processed to an appropriate particle size for mixing with the zeolite. The salt first undergoes size reduction in a commercially available rock crusher. It is then fed into a Prater impact mill with a classifier where it is reduced to an average particle size of 245 μm. As part of nonradioactive testing, over 180 experiments were performed by ANL with the mill/classifier to evaluate the capabilities of the equipment and determine particle size requirements.

The salt and zeolite are combined and processed in a V-mixer where they are heated to 500 °C for approximately 15 h. In this operation the salt is occluded into the zeolite structure. Twenty-two V-mixer experiments were performed by ANL with nonradioactive materials. The V-mixer is also used to mix 25 wt % glass with the salt-loaded zeolite. The nominal batch size is 50 kg, but an 80-kg batch was also successfully tested by ANL.

The glass/salt-loaded zeolite mixture is loaded into stainless-steel cans for processing through the HIP. Before the cans are sealed, they are evacuated. A tungsten inert gas weld is used to seal the cans. More than 100 demonstration-scale HIP cans were processed prior to operations with radioactive materials. Two can styles, one of which is pictured in Figure 3.13, were successfully tested through the demonstration. The can was provided as part of a cooperative research and development agreement with the Australian Nuclear Science and Technology Organisation (ANSTO). The result of this extensive out-of-cell testing was the development of operating parameters and acceptance criteria for each of the ceramic waste form process steps.

Once all the equipment was installed in the HFEF, operations with radioactive materials commenced using the parameters developed from nonradioactive testing. The salt used for these tests came from the Mark-IV electrorefiner after treatment of 100 driver assemblies. The salt was removed from the electrorefiner using a tray system lowered into an electrorefiner port. After the tray filled with molten salt, it was removed, and the salt

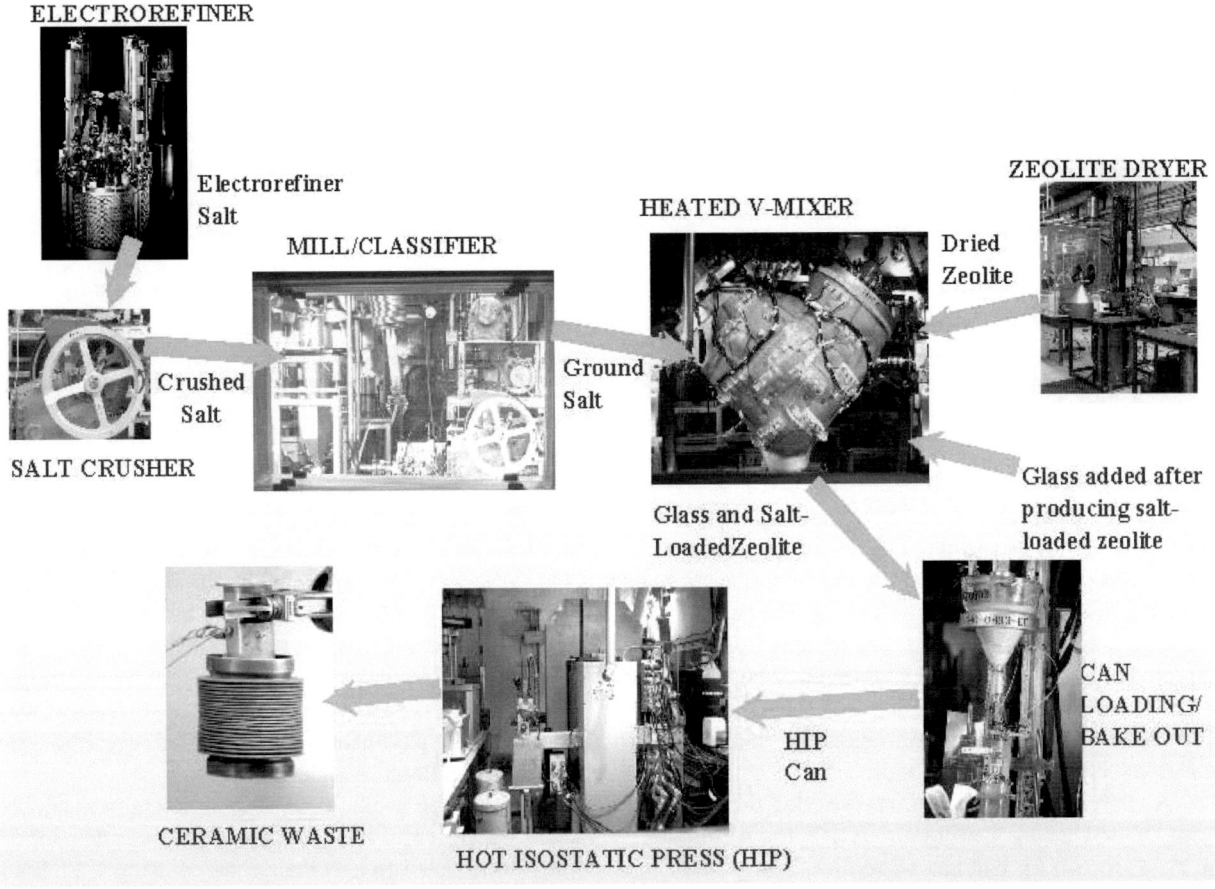

FIGURE 3.12 Ceramic waste process flow sheet.
SOURCE: Argonne National Laboratory.

solidified. The salt released easily from the tray. It was transferred by cask from the FCF to HFEF. It contained 9.4 wt % uranium chlorides, 0.5 wt % transuranic chlorides, and 5.2 wt % fission product chlorides. Four kilograms of salt were processed into a 50-kg batch of glass and salt-loaded zeolite. From this material, 10 HIP cans were processed through the HIP.

The cans used in the demonstration HIP are 4.5 in. in diameter (Figure 3.13). For inventory operations, the can diameter will be approximately 20 in. Scaling tests were performed using the nonradioactive materials. Three 8-in. diameter cans were first processed. The time at maximum temperature and the pressure in the HIP was increased from one hour for the demonstration-scale can to two hours for the 8-in. can. Characterization results indicated that the product in these cans had the same density, same phases present, and same performance based on leach tests as at the demonstration scale. Based on these results, two 18-in. diameter cans were processed. Characterization of these cans is ongoing. Initial X-ray diffraction results show that the same material phases are present in these cans as in cans of the other two sizes even though the hold time was 10 h.

In summary, the operating parameters for the production of the ceramic waste form have been developed and applied successfully to operations with radioactive materials at the demonstration scale. All the equipment for

FIGURE 3.13 Demonstration-scale hot isostatic pressure waste form can.
SOURCE: Argonne National Laboratory.

demonstration operations, except for the HIP, are close to the size required for inventory operations. Based on scaling tests with nonradioactive materials, the feasibility to scale HIP operations to production size has also been successfully demonstrated.

Pressureless Sintering

ANL is also investigating alternatives to use of the HIP for ceramic waste form fabrication. In particular, a process of pressureless sintering, or consolidation, is being studied that may have certain advantages over the HIP process.

The preparation steps for pressureless sintering are essentially identical to those for the HIP process. Zeolite A, glass binder, and waste-loaded salt are mixed in the V-mixer within the hot cell. The blended powder is poured into graphite molds ("setters") that are passed through a tunnel kiln at a fixed temperature and time. The processed waste form is then removed from the mold, allowing re-use of the mold, and transferred into a storage container until it is loaded into a canister for final disposal.

Studies are being conducted to establish optimal fabrication procedures and to determine whether pressureless sintering can produce a suitable waste form. The current reference processing conditions for the kiln are 850 °C for approximately 4 hours at 1 atmosphere pressure of argon gas. Initial studies indicate that a higher volume percentage (50 vol %) of inert glass binder is needed to fabricate a reproducible waste form. Preliminary analysis indicates that the waste form produced by pressureless sintering has a density equal to that of the waste form produced by the HIP process, within analytical uncertainties. The short-term leaching characteristics of both waste forms are also essentially identical, although the committee has previously noted that such short-term tests may not be fully indicative of long-term release-rate performance of waste forms under expected repository conditions.[12] Furthermore, the heterogeneous nature of the multiphase ceramic waste form mandates examination of the microstructure and phase composition of as-produced waste forms.

Pressureless sintering may provide advantages over the HIP process during fabrication. As noted in the committee's ninth report, the use of high-pressure HIP devices in hot cells may present unresolved safety and sustained-operation issues.[13] Compared to the HIP process, the pressureless sintering method may also give a safer and easier pathway to volumetric scale-up of waste form fabrication. Further demonstration is needed and is being pursued by ANL. Higher rates (mass/time) of fabrication may also be achievable by pressureless sintering for equivalent-size fabrication equipment, although whether waste form fabrication is a rate-limiting step in the overall electrometallurgical processing technique is not apparent.

Recommendation: Studies to compare the type, abundance, and radionuclide inventory of minor and trace phases between ceramic waste forms produced by pressureless sintering versus the HIP process should be given high priority in the post-demonstration phase.

COMMITTEE ANALYSIS OF ALTERNATIVE METHODS FOR TREATMENT OF DOE SNF

A variety of other methods, besides electrometallurgical technology, are either in use or have been proposed for the treatment of spent nuclear fuel. The committee has evaluated alternatives to electrometallurgical technology in two of its reports.[14,15] The following summarizes the committee's evaluations of these alternative technologies in light of technical advances in the development of electrometallurgical treatment of spent nuclear fuel.[16]

Direct Disposal

From a materials and design standpoint, overpack technologies appear to be technically sound. However, the long-term durability of the proposed overpack container has not been demonstrated or documented. Without such a demonstration of extended containment, the ability of the HIC disposal concept to meet the stated safety goals proposed by the National Research Council is unknown.[17] At the present time, direct emplacement of EBR-II SNF is precluded by DOE policy concerning acceptance of RCRA-designated mixed waste (which contains both hazardous and radioactive waste). Because of the presence of both metallic uranium and sodium, EBR-II SNF is categorized as an RCRA hazardous waste that is potentially both pyrophoric and reactive.

---

[12]National Research Council, *Electrometallurgical Techniques for DOE Spent Fuel Treatment: An Assessment of Waste Form Development and Characterization*, National Academy Press, Washington, D.C., 1999, p. 26.

[13]National Research Council, *Electrometallurgical Techniques for DOE Spent Fuel Treatment: An Assessment of Waste Form Development and Characterization*, National Academy Press, Washington, D.C., 1999, reference 42.

[14]National Research Council, *An Assessment of Continued R&D into an Electrometallurgical Approach for Treating DOE Spent Nuclear Fuel*, National Academy Press, Washington, D.C., 1995. In this report the committee considered spent fuel treatment alternatives to EMT within the context of all DOE SNF.

[15]National Research Council, *Electrometallurgical Techniques for DOE Spent Fuel Treatment: Spring 1998 Status Report on Argonne National Laboratory's R&D Activity*, National Academy Press, Washington, D.C., 1998. In this report the committee considered spent fuel treatment within the context of EBR-II SNF.

[16]This topic was examined in depth in the committee's seventh report: National Research Council, *Electrometallurgical Techniques for DOE Spent Fuel Treatment: Spring 1998 Status Report on Argonne National Laboratory's R&D Activity*, National Academy Press, Washington, D.C., 1998, pp. 13-21.

[17]National Research Council, *Technical Bases for Yucca Mountain Standards*, National Academy Press, Washington, D.C., 1995, pp. 71-72.

Glass Material Oxidation and Dissolution System (GMODS)

The basic concept of GMODS is to add unprocessed SNF and a sacrificial oxide to a glass melter at about 1000 °C, where a lead-borate glass acts as a solvent, and PbO oxidizes SNF metal components in situ to metal oxides which dissolve in the glass, and metallic lead. The lead metal (with dissolved metals) separates from the molten glass and sinks to the bottom of the melter. The GMODS process would be an attractive general approach that could be employed if it were successfully developed. However, considering the time period and cost necessary for development of the GMODS relative to the level of maturity of the EMT process, GMODS does not appear to be a viable alternative for processing only EBR-II SNF unless the process would also be applied to other DOE SNFs and miscellaneous fissile materials.

Melt and Dilute

The melt and dilute process chops SNF elements and melts them at approximately 650-850 °C, and then dilutes them by addition of depleted uranium and iron. No experimental work relevant to the processing of EBR-II SNF has been carried out. No development schedule or cost estimates were presented; therefore, the committee does not have sufficient information to evaluate melt and dilute as a viable alternative to EMT.

PUREX Process

The PUREX process is a counter-current solvent extraction method used to separate and purify uranium and plutonium from fission-product-containing SNF and irradiate uranium targets. The PUREX process is well developed and has been used to treat SNF and irradiated uranium for over 40 years. The development of a versatile head-end process to handle mechanical decladding, sodium removal, and zirconium sludge formation for EBR-II SNF for the SRS PUREX facility does not seem justified solely for the purpose of treating the relatively small quantity of EBR-II fuel that will remain after completion of the EMT demonstration. However, DOE would need to consider in the broader context whether developing a versatile head-end treatment step for the SRS PUREX facility is an attractive means for treating large quantities of DOE SNF (e.g., Zr or zircaloy-clad fuels such as N-reactor fuel) and other fissile-containing materials. A significant issue for treating EBR-II fuel at SRS by PUREX relates to public concerns about transportation of the fuel from the current storage site at ANL-W to SRS.[18] A final consideration is that at the present time the PUREX canyons at SRS are due to be shut down and would not be available for this work according to Administration Policy.

Chloride Volatility

The unit operations for chloride volatility conist of: (1) a high-temperature chlorination step that operates at approximately 1500 °C and converts metallic fuel and cladding materials to gaseous chloride compounds, (2) a molten zinc chloride bed that removes the TRU chlorides and most of the fission products and operates at approximately 400 °C, (3) a series of fluidized beds and condensers operating at successively lower temperatures to condense zirconium tetrachloride, uranium hexachloride, and stannous tetrachloride, and (4) a zinc chloride regeneration/recycle process. The TRU and fission product chloride are then converted to either fluorides or oxides for final disposal. As proposed, this technology is applicable to a very limited range of fuel types. A reducing agent, such as carbon monoxide, would be needed in the chlorination step to prevent the formation of oxychloride compounds even for metallic fuel forms. The behavior of some of the constituents of EBR-II fuel, such as metallic sodium, may also limit the suitability of this process due to the accumulation of considerable amounts of sodium chloride during the chlorination step. Other unknown quantities, including the behavior of

---

[18] For a discussion of transportation issues regarding spent fuel and other nuclear wastes, see National Research Council, *Nuclear Wastes Technologies for Separations and Transmutation*, National Academy Press, Washington, D.C., 1996, pp. 102-103.

stainless steel, in the chlorination step further decrease the potential applicability of this process to stainless steel-clad SNF (e.g., EBR-II fuel). At present, considering also cost and schedule, the chloride volatility process is not competitive with the current EMT process. In addition, the committee is aware that a significant amount of work on chloride volatility processes was conducted during the 1960s at ANL and ORNL.[19] However, this earlier work does not suggest to the committee that this approach is an attractive alternative to EMT.

Plasma Arc

In this process, components of SNFs are melted and oxidized, with the help of an oxygen lance, in a rotating furnace containing molten ceramic materials at a temperature of 1600 °C or higher. Although plasma arc processing has been used successfully to treat nonradioactive and low-level radioactive wastes, significant research, development, and demonstration would be needed to process SNF because of the much higher fission product and fissile material content. With regard specifically to EBR-II fuels, extensive work would have to be performed to ensure that the plasma arc process would be compatible with safely processing potentially pyrophoric uranium, and volatile and reactive sodium metal. Unresolved safety issues at present preclude consideration of plasma arc processing as a viable alternative to the EMT process.

---

[19]T.A. Gens, "Chloride Volatility Processing of Nuclear Fuels," *Chem. Eng. Prog. Symp. Ser., Nuclear Engineering*, Part X (47), Vol. 60, 1964, pp. 37-47.

4

Waste Streams Produced by the Electrometallurgical Technology Process

The electrometallurgical treatment of spent EBR-II reactor fuel involves a set of operations designed to disassemble driver and blanket fuel pins, to refine and recover the uranium metal contained therein, and to segregate the radioactive waste components. The radioactive waste components are consolidated into two forms: the ceramic waste form (CWF) includes actinide elements and fission products in a glass-ceramic matrix; the metal waste form (MWF) contains noble metal fission products in a fuel-cladding matrix.[1]

Both the MWF and the CWF are high-level waste (HLW) forms intended for final disposition in a geologic repository. The fate of the recovered uranium stream has not yet been decided, although a minor waste stream, the dross from the cathode processor and casting furnace operations, would be disposed of as TRU waste. The mass balance for all these output streams is shown in Table 4.1 for 84 driver fuel assemblies that were processed by ANL.

WASTE FORM QUALIFICATION

Argonne National Laboratory's criterion 2 (Chapter 6) for evaluating the success of EBR-II spent nuclear fuel demonstration project addresses the development and determination of performance characteristics of the waste forms produced in the EMT process. Performance relates specifically to acceptable levels of long-term (10,000 years or longer) release of radionuclides from waste forms placed in a deep geologic repository.

The Department of Energy (DOE), through its Office of Civilian Radioactive Waste Management (OCRWM/RW), and in conjunction with the development of final waste acceptance criteria to be based on Environmental Protection Agency/Nuclear Regulatory Commission Agency regulations, is currently assessing the viability of permanent disposal of spent nuclear fuel (SNF) in a deep geologic repository at Yucca Mountain, NV.[2] Long-term isolation from the biosphere of nuclear materials in the repository is to be ensured through a defense in depth based on the use of multiple natural and engineered barriers. The performance and compatibility of the ANL waste

---

[1] Much of the material in this chapter is presented in the committee's ninth report, National Research Council, *Electrometallurgical Techniques for DOE Spent Fuel Treatment: An Assessment of Waste Form Development and Characterization*, National Academy Press, Washington, D.C., 1999.

[2] Office of Civilian Radioactive Waste Management, *Viability Assessment of a Repository at Yucca Mountain*, U.S. Department of Energy, DOE/RW-0508, Washington, D.C., 1998.

TABLE 4.1 Material Balance Estimates (g) for Selected Components of EBR-II Driver Fuel[a]

	U	Pu	Np	Na	^{144}Ce	^{137}Cs	^{136}Ru	^{125}Sb	Tc
Input									
Electrorefiner feed	427×10^3	1509	156	9801	26	1031	5	4	757
CP/CF feed[b]	610×10^3	0	0	1	0	0	0	0	0
Total input	1037×10^3	1509	156	9802	26	1031	5	4	757
Output									
MWF[c]	25×10^3	165	12	723	2	62	4	2	660
Uranium[d]	955×10^3	4	3	3	0	0	0	0	2
Dross[e]	14×10^3	0	0	0	0	0	0	0	0
CWF[f]	38×10^3	1248	124	7894	17	722	0	0	0
Other[g]	8×10^3	5	2	309	0.1	5	0	0	4
Total output	1039×10^3	1422	140	8929	19	789	4	2	666

[a] SOURCE: R.D. Mariani, D. Vaden, B.R. Westphal, D.V. Laug, S.S. Cunningham, S.X. Li, T.A. Johnson, J.R. Krsul, and M.J. Lambregts, *Process Description For Driver Fuel Treatment Operations*, ANL Technical Memorandum No. 11, Argonne National Laboratory, Argonne, IL, Table 23, p. 35.
[b] Depleted uranium added in cathode processor (CP)/casting furnace (CF) operations.
[c] Cladding hulls (to be converted to metal waste form, MWF).
[d] Uranium ingots for interim storage.
[e] Dross from cathode processor (CP)/casting furnace (CF) operations.
[f] Material remaining in electrorefiner salt (to be converted into ceramic waste form CWF).
[g] Material remaining in electrorefiner holdup, cadmium pool, and plenum sections.

forms must therefore be assessed within this system context of overall repository safety. The committee recognizes that it is not waste form performance per se, but rather the safe performance of the integrated system of engineered and natural barriers, that must be demonstrated. It is within this system context that the committee has evaluated ANL's progress toward obtaining the necessary DOE-RW acceptance of EMT waste forms, qualifying them for final disposal in a geologic repository in the future.

To date, both commercial SNF and vitrified defense HLW have been subjected to detailed characterizations of their radionuclide-release performance under expected repository conditions.[3] The resulting data have been used to guide isolation strategies. These data also have led to an initial design of engineered barrier systems (EBS) and in the assessment of safety and viability of waste forms for placement in a geologic repository.[4]

Acceptance of DOE SNF and HLW waste forms by DOE-RW for final geological disposal involves many characterization and testing issues with respect to quality assurance and performance. To ensure a coordinated effort between DOE-RW and DOE's Office of Environmental Management (DOE-EM) to resolve such issues, a memorandum of agreement (MOA) was issued[5] that establishes the terms and conditions under which DOE-RW will permit DOE-EM to dispose of its SNF and HLW. The MOA identifies the responsibilities of DOE-RW and DOE-EM for data on transportation, storage (if needed), safeguards, characterization, and final acceptance for disposal of these materials. Responsibility to treat the EBR fuel rests with DOE-NE, but the ultimate disposition of this treated EBR-II fuel and any HLW generated is the responsibility of DOE-EM. Hence, the committee presumes that the waste qualification activities of the EMT program will be guided and governed in the post-demonstration period by the MOA.

---

[3] U.S. Department of Energy, *Mined Geologic Disposal System Waste Acceptance Criteria Document*, B00000000-01717-4600-00095 REV 00, TRW Environmental Safety Systems, Inc., Las Vegas, NV, 1997, pp. 5-1 – 5-8.

[4] U.S. Department of Energy, *Mined Geologic Disposal System Waste Acceptance Criteria Document*, B00000000-01717-4600-00095 REV 00, TRW Environmental Safety Systems, Inc., Las Vegas, NV, 1997.

[5] U.S. Department of Energy, *Memorandum of Agreement for Acceptance of Department of Energy Spent Nuclear Fuel and High-Level Radioactive Waste between the Assistant Secretary for Environmental Management (EM) U.S. Department of Energy (DOE) and the Director, Office of Civilian Radioactive Waste Management (RW) U. S. DOE*, Washington, D.C., September 1998.

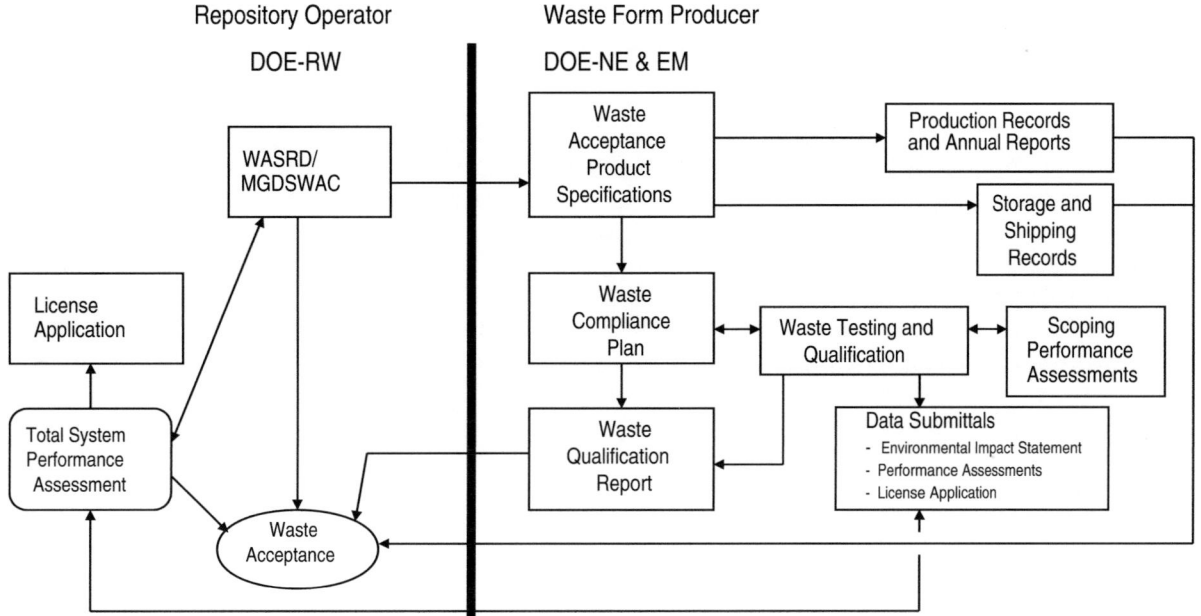

FIGURE 4.1 Waste qualification flow diagram.
SOURCE: Argonne National Laboratory.

Figure 4.1, reprinted from ANL's report on its waste form qualification strategy,[6] is a flow diagram of the interrelated waste characterization/verification activities of both DOE-RW and the producers of DOE HLW (DOE-EM and DOE-NE).[7] Based on these activities, documents are to be produced that will support the decision by DOE-RW on acceptance of EMT HLW waste forms. ANL's flow diagram is based on similar flow diagrams presented in "Appendix C: Sub-agreement on the DOE SNF and HLW Technical Baseline" in the MOA and in the Savannah River Laboratory Defense Glass Program Manual.[8] In particular, Appendix C of the MOA addresses issues related to the development, concurrence, distribution, compliance, and conformance verification of acceptance criteria for DOE SNF and HLW.

> **Finding:** ANL has adopted a waste qualification strategy that is appropriately based on guidance provided in the MOA between DOE-RW and DOE-EM regarding acceptance of DOE SNF and HLW.

---

[6]T.P. O'Holleran, R.W. Benedict, and S.G. Johnson, *Waste Form Qualification Strategy for the Metal and Ceramic Waste Forms from Electrometallurgical Treatment of Spent Nuclear Fuel*, NT Technical Memorandum No. 115, Argonne National Laboratory, Argonne, IL, 1999.

[7]Presented by Robert W. Benedict to the committee, National Academies Beckman Center, Irvine, CA, January 28, 1999.

[8]Westinghouse Savannah River Company, *DWPF Waste Acceptance Reference Manual (U)*, WSRC-IM-93-45, Westinghouse Savannah River Company, Aiken, SC, 1993.

The repository operator's (DOE-RW) specific waste qualification requirements for waste form producers (DOE-NE and DOE-EM) are shown in Figure 4.1. The key terms are defined in Appendix C of the MOA[9] as follows:

- *Acceptance*: The transfer of responsibility, custody, and physical possession of DOE SNF or HLW from DOE-EM to DOE-RW at the DOE-EM site.
- *Acceptance Criteria*: All technical and programmatic requirements that must be satisfied by DOE SNF and HLW for the repository program to meet regulatory requirements. DOE-RW is currently preparing a "Civilian Radioactive Waste Management System (CRWMS) Acceptance Criteria" document.
- *Waste acceptance product specifications (WAPS)*: The documentation by a producer of HLW that identifies the technical specifications for the HLW waste forms.
- *Waste form compliance plan*: The documentation prepared by a producer of HLW that describes planned analyses, tests, and engineering development work to be undertaken, as well as information to be included in waste form production records to demonstrate compliance of the proposed waste form with waste acceptance specifications.
- *Waste form qualification report*: The documentation prepared by a producer of HLW that describes results of analyses, tests, and engineering development work performed to demonstrate waste form compliance with acceptance specifications.

Waste Acceptance Product Specifications

WAPS data support quality assurance by enabling verification that the as-produced HLW waste forms conform consistently to acceptance specifications for disposal. Characteristics considered/examined include particulate size, pyrophoricity, dimensions, major phase chemistry, radionuclide inventories, heat-generation rate, phases in which the radionuclides are located, particle size, and solubility. These characteristics correspond to the "Waste Acceptance Product Specifications" of Figure 4.1.

The early phase of collecting WAPS data on EMT-produced waste forms was conducted during the demonstration project and is described in following sections of Chapter 4 on EMT waste forms. Data collection during the demonstration project was purposefully directed to provide data supporting issuance of an environmental impact statement (EIS) regarding continued application of the EMT process to DOE's remaining inventory of EBR-II spent fuel. Thus, ANL has oriented its current activities to provide evidence of successful compliance with demonstration criteria (see Chapter 6). To support a final waste-acceptance decision by DOE-RW, however, considerably more qualification and characterization data, especially on fully loaded radioactive EMT waste forms, will have to be collected during the post-demonstration period (see Chapter 5).

Finding: At the end of the demonstration project in late 1999, WAPS were developed based on non-radioactive and radioactive EMT waste forms. Preliminary testing and modeling of the performance of EMT waste forms under repository conditions were also initiated during the demonstration project.

ANL's WAPS[10,11] are patterned after the quality assurance protocols used for Defense Program borosilicate glass (DHLW) HLW.[12] The committee observes, however, that DHLW borosilicate glass has *not* received final

---

[9]U.S. Department of Energy, *Mined Geologic Disposal System Waste Acceptance Criteria Document*, B00000000-01717-4600-00095 Rev. 00, TRW Environmental Safety Systems, Inc., Las Vegas, NV, 1997, Appendix C.

[10]T.P. O'Holleran, R.W. Benedict, and S.G. Johnson, *Waste Form Qualification Strategy for the Metal and Ceramic Waste Forms from Electrometallurgical Treatment of Spent Nuclear Fuel*, NT Technical Memorandum No. 115, Argonne National Laboratory, Argonne, IL, 1999.

[11]T.P. O'Holleran, D.P. Abraham, J.P. Ackerman, K.M. Goff, S.G. Johnson, and D.D. Keiser, *Waste Acceptance Product Specifications for the Waste Forms from Electrometallurgical Treatment of Spent Nuclear Fuel*, NT Technical Memorandum No. 116, Argonne National Laboratory, Argonne, IL, 1999.

[12]Westinghouse Savannah River Company, DWPF Waste Acceptance Reference Manual (U), WSRC-IM-93-45, Westinghouse Savannah River Company, Aiken, SC, 1993.

qualification and acceptance by DOE-RW for geologic disposal. Furthermore, the committee has reservations about whether the DHLW WAPS protocols, developed for a homogeneous, single-phase waste form, can be directly and without significant modification used as the basis for the WAPS of heterogeneous, multiphase EMT waste forms. ANL has recognized the need to refine its current test protocols to address the multiphasic nature of the ceramic waste form. Characterization of the distribution of radionuclides among coexisting phases is but one item that would seem to be necessary as a quality-assurance specification for any multiphase HLW waste form, similar to information required for SNF.[13,14]

Waste Form Acceptance Criteria

Waste testing and qualification involves data more directly connected to the long-term (10,000 years or greater) performance characteristics of HLW waste forms under expected repository conditions. The committee assumes that a description of the rationale and methodology for collecting such data will be part of the acceptance criteria that have yet to be finalized by the DOE-RW. Section VII "Acceptance Criteria" of the MOA between DOE-RW and DOE-EM noted that DOE-RW would be responsible for the characterization of long-term performance of HLW starting in Fiscal Year 1999.[15]

Finding: The committee understands that the DOE is preparing waste acceptance criteria, including guidance on long-term waste form performance testing and qualification. This new document may modify the actual waste acceptance strategies and waste acceptance criteria that the EMT program is currently following.

A fuller examination of waste form acceptance criteria based on long-term performance evaluation by ANL extending beyond the demonstration project is made in Chapter 5 (Waste Form Qualification from a Repository Perspective, Potential for Alternative Nontesting Strategies for Waste Acceptance).

METAL WASTE FORMS

Background

The MWF may contain up to 4 wt % noble metal fission products and up to 11 wt % uranium.[16,17] The long-term corrosion behavior of this type of alloy is not known and must therefore be determined in the corrosion tests carried out at Argonne National Laboratory-East (ANL-E). The EBR-II driver fuel is primarily a uranium-10 wt % zirconium alloy with types 316, D9, or HT9 stainless steel (SS) cladding. Because zirconium is a principal component of the driver fuel, zirconium will be a significant component of the metal waste stream. For the entire EBR-II spent fuel inventory, the base metal waste stream composition is stainless steel containing approximately 15 wt % zirconium, labeled SS-15Zr by ANL personnel.[18] Thus, MWF testing at ANL-W has focused primarily on this and similar alloys.

---

[13]Office of Civilian Radioactive Waste Management, *Total System Performance Assessment for Viability Assessment*, DOE/RW-0508, U.S. Department of Energy, Washington, D.C., 1997, Chapter 6.

[14]L. Johnson and J. Tait, *Release of Segregated Nuclides from Spent Fuel*, SKB Technical Report 97-18, Swedish Nuclear Fuel and Waste Management Co., Stockholm, Sweden, 1997.

[15]U.S. Department of Energy, *Memorandum of Agreement for Acceptance of Department of Energy Spent Nuclear Fuel and High-Level Radioactive Waste between the Assistant Secretary for Environmental Management (EM) U.S. Department of Energy (DOE) and the Director Office of Civilian Radioactive Waste Management (RW) U. S. DOE*, Washington, D.C., September 1998.

[16]Material balance for the MWF is given in Table 4.1.

[17]For a listing of the isotopic composition of the metal waste form for blended fuel, see K.M. Goff, L.L. Briggs, R.W. Benedict, J.R. Liaw, M.F. Simpson, E.E. Feldman, R.A. Uras, H.E. Bliss, A.M. Yacout, D.D. Keiser, K.C. Marsden, and C.W. Nielsen, *Production Operations for the Electrometallurgical Treatment of Sodium-Bonded Spent Nuclear Fuel*, NT Technical Memorandum No. 107, Argonne National Laboratory, Argonne, IL, 1999, Appendix B.

[18]D.P. Abraham, *Metal Waste Form Handbook,* NT Technical Memorandum No. 88, Argonne National Laboratory, Argonne, IL, 1998, p. 3.

Metal Waste Form Testing and Qualification Tests

The MWF is obtained by melting at 1600 °C in an inert atmosphere the cladding residue that remains from the electrorefiner step. The molten residue is adjusted to contain 15 wt % zirconium and then cast into ingots. Corrosion resistance and noble metal fission product retention are the principal requirements for the safe application of the MWF. Waste-form qualification involves experimental testing and model development. Research at ANL-E has evaluated alloy metallurgy and alloy properties, including mechanical and thermophysical properties and corrosion behavior.[19] The corrosion resistance of SS-15Zr alloys has been determined using immersion tests, electrochemical tests, and accelerated corrosion tests (vapor hydration, high-temperature immersion, and product consistency tests).

Plans for post-demonstration qualification testing beyond June 1999 and testing highlights have been presented at several committee meetings.[20,21] The MWF test plan consists of attribute tests, characterization tests, accelerated tests, and service condition tests. The attribute tests, as defined by ANL, are designed to provide material property information using electron microscopy, x-ray analysis, and neutron diffraction. Good progress seems to have been achieved in the identification of the various phases of SS-15 Zr-type materials. Noble metal-rich precipitates have not been observed. The physical properties of SS-15Zr, a very strong alloy with mechanical and thermophysical properties comparable to those of other alloys, indicate that this material is suitable as a waste form material.

The characterization tests consist of immersion testing in sealed Teflon vessels at 90 °C in J-13 (simulated Yucca Mountain well water) and deionized water. The tests that have been terminated showed either no corrosive attack or only minor tarnish. The test solutions have been submitted for elemental analysis. ANL planned a total of 856 tests as necessary to achieve the goals of the demonstration project. In the view of the committee this seems an excessive number.

The accelerated tests are designed to shorten the test period and consist of immersion in deionized water in a titanium vessel at 200 °C for 28 days. Six alloy compositions were tested. Corrosion rates were very low and no correlation of elemental leaching with alloy composition was found.[22]

Electrochemical corrosion testing, based on the polarization resistance technique (ASTM G59)[23] for measuring instantaneous corrosion rates,[24] is used to screen out alloy compositions that may not be suitable for repository disposal. An example was given[25] for a stainless steel alloy, SS-1Ag-2Nb-1Pd-1Ru, that did not contain zirconium and had high corrosion rates at pH 2. Corrosion rates were low for alloys that contained from 15 to 20 wt % zirconium. Corrosion rates were low for the MWF alloys in J-13 and in solutions at pH = 2, 4, and 10 and were similar to those for SS316 and alloy C22. These results are not surprising considering that the solutions tested did not contain chloride ions that could have initiated localized corrosion. Corrosion rate data for MWF materials were also compared to those of copper and mild steel.

The results from pulsed-flow immersion tests of SS-15Zr alloys containing Nb, Pd, Rh, Ru, and Tc in J-13 water at 90 °C for up to 275 days showed a sudden increase in Tc release rates after about 150 days; however, the overall release rate remained relatively small. The cause of this behavior is under investigation. The results of the immersion tests, which have shown that only small amounts of fission and actinides are dissolved in the test solution, suggest that corrosion of the SS-15Zr MWF is not a dominant release mechanism. Corrosion appears to be retarded by the formation of a passivating oxide layer, typical for austenitic stainless steels, that may trap the

---

[19] Presentation by Stephen G. Johnson and Daniel Abraham to the committee, ANL-W, June 25-26, 1998.

[20] Presentation by Stephen G. Johnson and Daniel Abraham to the committee, ANL-W, June 25-26, 1998.

[21] Presentation by Daniel Abraham to the committee, ANL-E, October 26-27, 1998.

[22] Presentation by Daniel Abraham to the committee, ANL-W, June 21, 1999.

[23] ASTM G59-91, "Standard Method for Conducting Potentiodynamic Polarization Resistance Measurements," American Society for Testing and Materials, West Conshohocken, PA, 1991.

[24] F. Mansfeld, "The Polarization Resistance Method for Measuring Corrosion Currents," in *Adv. Corros. Sci. Technol.*, Vol. 6, 1976, p. 163.

[25] Presentation by Daniel Abraham to the committee, ANL-W, June 21, 1999.

fission products and actinides, limiting their release. Tests have not yet been completed with added uranium. The committee recommended in its ninth report that surface analyses by x-ray photoelectron spectroscopy (XPS) or Auger electron spectroscopy (AES) of the passive layers formed on the MWF samples should be performed in order to compare the chemical composition of these films with those formed on SS 304 or SS 316.[26]

Finding: Some of the corrosion products, which may sequester radionuclides, might remain on the sample surface and might therefore not be detected by solution analysis.

Recommendation: Surface analysis by X-ray photoelectron spectroscopy (XPS) or Auger electron spectroscopy (AES) should be continued in the post-demonstration phase for selected samples to determine the chemical composition of passivating films and/or corrosion products.

Galvanic corrosion tests according to ASTM G71 have indicated that enhanced corrosion of SS-15Zr due to galvanic coupling of the MWF with the inner lining of the waste form container (assumed to be alloy C22) is not likely to be significant.[27] Both materials were in the passive state in the test solution, and the SS-15Zr was electrochemically noble to alloy C22; i.e., it was cathodically protected.

Vapor hydration tests providing a direct assessment of the long-term durability of the MWF have been performed in sealed SS vessels for 56 and 182 days in superheated steam at 200 °C.[28] In initial tests it was found that corrosion rates were greatly accelerated by exposure to steam. Samples containing less than 5 wt % zirconium (or no zirconium) were heavily rusted and contained numerous pits. The corrosion product layer for SS-15Zr has been estimated to have a thickness of less than 1 μm after 182 days, based on visual observations. The chemical nature of the corrosion products is under investigation. Further tests at 200 °C and 100% relative humidity for 56 days showed that corrosion damage of the standard MWF was negligible, while severe corrosion occurred for pure Cu and Fe. Apparently an adherent Cr-rich oxide layer, which will be examined in more detail by TEM and AES in the post-demonstration period, retarded the corrosion reaction.

ANL personnel discussed corrosion testing of SS-15Zr MWF samples at a 1998 meeting, concluding that "SS-Zr waste forms are very resistant to the normal corrosion conditions envisioned at the proposed Yucca Mountain geologic repository."[29] The effect of radiation on corrosion behavior has been discussed only briefly in presentations to the committee by ANL personnel.[30] Calculations carried out at ANL seem to suggest that radiation levels in the MWF are too low to affect the corrosion resistance.

The toxicity characteristic leaching procedure (TCLP) test data suggest that the MWF passes the TCLP test.[31] The results for the release of Ag, As, Ba, Cd, Cr, Hg, Pb, and Se were below the detection limits of acceptable methods.

Based on the results from the various corrosion tests, ANL personnel concluded that SS-15Zr shows corrosion behavior similar to that of austenitic stainless steels such as SS316.[32] High corrosion rates were observed in electrochemical and vapor hydration tests for alloys with less than 5 wt % zirconium. The corrosion resistance of the MWF appears to be dominated by the passivation behavior of the alloy, and dissolution is not considered to be a dominant mechanism for release of the radionuclides.

---

[26]National Research Council, *Electrometallurgical Techniques for DOE Spent Fuel Treatment: An Assessment of Waste Form Development and Characterization*, National Academy Press, Washington, D.C., 1999.

[27]Presentation by Daniel Abraham to the committee, ANL-W, June 21, 1999.

[28]Presentation by Dennis D. Keiser, Jr. to the committee, ANL-W, June 21, 1999.

[29]D.P. Abraham, L.J. Simpson, M.J. Devries, and S.M. McDeavitt, "Corrosion Testing of Stainless Steel-Zirconium Metal Waste Forms," *Scientific Basis for Waste Management XXII*, Materials Research Society, Boston, 1998.

[30]Presentation by Daniel Abraham to the committee, ANL-W, June 21, 1999.

[31]Presentation by Dennis D. Keiser, Jr., to the committee, ANL-W, June 21, 1999.

[32]Presentations by Daniel Abraham and Dennis D. Keiser, Jr., to the committee, ANL-W, June 21, 1999.

Finding: Results from corrosion testing of the MWF in rather benign environments suggest that the corrosion behavior of the MWF is similar to that of austenitic stainless steel.

The tests to be performed after June 1999 have not been finalized. Electrochemical tests are to be performed at elevated temperatures in order to shorten the test period. It was suggested by the committee that ANL concentrate on a few key samples, expose them at higher temperatures and chloride concentrations, and obtain electrochemical and surface analysis data.[33] Tests are also to be conducted in chloride solutions with concentrations of up to 10,000 ppm, which are credible conditions that might be encountered in the repository.

Finding: ANL has carried out a number of corrosion tests using mild solutions. Under these conditions, significant corrosion damage to the MWF is not expected.

Recommendation: In the post-demonstration phase ANL personnel should subject a few carefully selected samples to additional evaluation by surface analysis to determine the chemical composition of the corrosion products.

Recommendation: ANL personnel should concentrate on a few key samples, expose them at higher temperatures and chloride concentrations, and obtain electrochemical and surface analysis data.

Guidance for carrying out pitting scans that provide a more direct measure of susceptibility to localized corrosion can be obtained from ASTM G61.[34] Corrosion rate data could also be obtained from such measurements. The proposed study of crevice corrosion needs careful design of an artificial crevice with consideration of the proposed application of these alloys. ASTM G48,[35] which describes a multiple crevice assembly, could serve as guidance. The use of an automated crevice device is being evaluated at ANL-E.

Modeling

Waste form degradation/radionuclide release models have been established that are an integral part of ANL's waste form repository performance assessment effort and will be used for predicting the long-term corrosion behavior of the MWF. The difficulties in ANL's modeling effort are the necessity to extrapolate experimental short-term data to extremely long time scales and the limitations imposed by the assumption that the corrosion mechanism(s) observed in the short-term tests will prevail in the long term, perhaps up to 10,000 years. The corrosion tests carried out so far are being used to reduce uncertainties in the models and to obtain an understanding of the relationships between alloy chemistry, microstructure, and the corrosivity of the test environment. Progress in this direction has been slow mainly because the MWF has proved very corrosion resistant in the rather benign test solutions used so far. It might be fruitful in future modeling efforts to use the information available for the long-term corrosion behavior of stainless steels. Important new information will become available from the crevice and pitting corrosion tests that are planned for the post-demonstration period.

---

[33] National Research Council, *Electrometallurgical Techniques for DOE Spent Fuel Treatment: An Assessment of Waste Form Development and Characterization*, National Academy Press, Washington, D.C., 1999, pp. 19-20.

[34] ASTM G61-86, "Standard Test Method for Conducting Cyclic Potentiodynamic Polarization Measurements for Localized Corrosion Susceptibility of Iron-, Nickel-, or Cobalt-Based Alloys," American Society for Testing and Materials, West Conshohocken, PA, 1986.

[35] ASTM G48-97, "Standard Test Method for Pitting and Crevice Corrosion Resistance of Stainless Steels and Related Alloys by Use of Ferric Chloride Solution," American Society for Testing and Materials, West Conshohocken, PA, 1997.

CERAMIC WASTE FORM

Background

As described in Chapters 2 and 3, the reference CWF for the demonstration was glass-bonded sodalite formed during the hot isostatic press (HIP) process. Sodalite ($Na_4Al_3(SiO_4)_3Cl$) is the thermolysis product formed from the salt-loaded zeolite 4A. In the electrometallurgical process (see Figure 2.1), the salt from the electorefiner containing TRUs and fission products is blended with dried zeolite in the V-mixer and is heated to occlude the salt into the zeolite. The salt-loaded zeolite is then densified into the CWF by HIP processing. By appropriate choice of temperature and pressure, the zeolite is converted into sodalite during HIP processing.

Characterization Tests

To support repository qualification of the CWF, ANL developed a protocol and conducted a variety of tests and analyses to provide the following information on the CWF: characterization of the phase distribution; waste form corrosion and radionuclide release rates; dissolution data pertinent to modeling CWF corrosion in total system performance assessment calculations; and development of a method that can be used to monitor product consistency and define working ranges for processing variables. Tests have been carried out on samples prepared at a laboratory scale with surrogate fission products, radioactive materials spiked with plutonium, and CWF produced with fully radioactive electrorefiner salt, from a 50 g to 10 kg scale. Detailed results and conclusions are contained in ANL's *Ceramic Waste Form Handbook*.[36] ANL's major findings are summarized here. More details are to be found in Appendices B and C and references cited therein.

Tests to determine CWF composition and structure include X-ray diffraction (XRD), scanning electron microscopy (SEM), transition electron microscopy (TEM), X-ray absorption fine structure spectroscopy, optical microscopy, and chemical analysis. Based on results from these tests the CWF can be characterized as a composite consisting of approximately 75% salt-loaded sodalite, 25% borosilicate glass, and up to about 5% of other minor phases, e.g., aluminosilicates, rare-earth silicates, oxides, and halite (NaCl).[37] There are only minor variations in the microstructure and composition among the laboratory-produced reference CWF containing surrogate fission products, CWF containing uranium and plutonium, and CWF produced from fully radioactive electrorefiner salt. The microstructure consists of polycrystalline sodalite grains comparable in size to the initial zeolite 4A grain size, encapsulated in intergranular glass. The minor component actinides and rare-earths form phases separate from the sodalite and glass phases. The actinides occur as nano-size crystal inclusions associated with the glass phase and/or near the glass/sodalite grain boundaries. Uranium and plutonium are found as oxides, oxide solid solutions, and silicate phases. Rare-earth fission products are found as silicates or mixed oxides that may contain actinides. Cesium and strontium appear to be evenly distributed throughout the waste form.

Accelerated alpha damage testing is being carried out on simulated CWF doped with 0.2 to 2.5 wt % ^{238}Pu or ^{239}Pu. The ^{238}Pu is shorter lived, and hence generates a larger alpha flux than would be found in the CWF.[38] The plutonium is observed primarily as oxide crystal inclusions in the intergranular glassy regions, ranging in size from

---

[36] W.L. Ebert, D.W. Esh, S.M. Frank, K.M. Goff, M.C. Hash, S.G. Johnson, M.A. Lewis, L.R. Morss, T.L. Moschetti, T.P. O'Holleran, M.K. Richmann, W. P. Riley, Jr., L.J. Simpson, W. Sinkler, M.L. Stanley, C.D. Tatko, D.J. Wronkiewicz, J.P. Ackerman, K.A. Arbesman, K.J. Bateman, T.J. Battisti, D.G. Cummings, T. DiSanto, M.L. Gougar, K.L. Hirsche, S.E. Kaps, L. Leibowitz, J.S. Luo, M. Noy, H. Retzer, M.F. Simpson, D. Sun, A.R. Warren, and V.N. Zyryanov, *Ceramic Waste Form Handbook*, NT Technical Memorandum No. 119, Argonne National Laboratory, Argonne, IL, 1999.

[37] For a listing of the isotopic composition of the ceramic waste form for blended fuel, see K.M. Goff, L.L. Briggs, R.W. Benedict, J.R. Liaw, M.F. Simpson, E.E. Feldman, R.A. Uras, H.E. Bliss, A.M. Yacout, D.D. Keiser, K.C. Marsden, and C.W. Nielsen, *Production Operations for the Electrometallurgical Treatment of Sodium-Bonded Spent Nuclear Fuel*, NT Technical Memorandum No. 107, Argonne National Laboratory, Argonne, IL, 1999, Appendix A.

[38] S.M. Frank, D.W. Esh, S.G. Johnson, M. Noy, and T.P. O'Holleran, "Effects of Alpha Decay Damage on the Structure and Leaching Rates of a Glass-Bonded Ceramic High Level Waste Form," *Conference Proceedings, Material Research Society, Symposium: Scientific Basis for Waste Management XXII*, Fall Meeting, Boston, Massachusetts, November 30 – December 4, 1998.

submicron to 20 microns in diameter. There is little or no plutonium dissolved in the glass or occluded in the sodalite. The PuO_2 phase shows an expected unit cell expansion due to alpha decay damage but as of the end of the demonstration project no bulk sample swelling had been observed. Initial results from this ongoing study show no significant degradation of the waste after 6 months at relatively low doses. Results are from x-ray diffraction lattice parameters of sodalite and bulk product consistency tests, and some SEM and density examination.

Elemental chemical analyses carried out on reference CWF and uranium-doped CWF show excellent agreement between calculated and measured compositions, verifying the validity of the analytical techniques.

Finding: The CWF is a multiphase, nonhomogeneous composite consisting of approximately 75% sodalite, 25% borosilicate glass, and up to 5% other minor phases, e.g., aluminosilicates, rare earth silicates, oxides, and halite (NaCl). The CWF waste form qualification program is based on adaptation of the models and test protocols developed for DHLW borosilicate glass.

Recommendation: The electrometallurgical technology program should continue to investigate and evaluate in the post-demonstration period whether the test protocols and conceptual models developed for monolithic single-phase borosilicate glass can adequately represent the behavior of the nonhomogeneous multiphase CWF.

Corrosion Tests

A variety of tests that monitor corrosion behavior were conducted by ANL to achieve a basic understanding of the processes that control dissolution of the CWF. The approach was based on the philosophy outlined in the ASTM C1174-98.[39] The basic steps for determining the mechanistic model are as follows:

1. Identify important alteration modes of the materials and important bounding disposal conditions.
2. Conduct tests to characterize alteration of the material under anticipated conditions that accelerate particular chemical or physical processes.
3. Develop a conceptual model for each alteration mode and measure values of model parameters.
4. Confirm the model with tests different from those used to provide the model parameter values.

A large number of corrosion tests have been designed and conducted by ANL to support CWF qualification. These scoping tests include studies of solution exchange with the CWF; product consistency tests (PCTs) in which the waste form is leached after crushing and sieving to achieve suitable particle sizes and washed to remove fines; MCC-1, a static leach test that uses a monolithic sample; pH stat tests; accessible-free-salt measurements; and vapor hydration. The MCC-1 test determines corrosion behavior under dilute solution conditions. Short-term MCC-1 tests measure dissolution behavior far from saturation, while long-term MCC-1 tests measure the approach to saturation. The PCTs reflect somewhat longer time dissolution and represent more repository-relevant conditions. None of these tests, however, reflects the open-system, mass-transport conditions that govern the actual release rates of radionuclides from the waste package system for disposal. The tests do provide data for assessment of the CWF's dissolution behavior and for use as input for a model to predict waste form behavior in the repository.

Tests have shown that the mechanism for release of radionuclides from sodalite is dissolution rather than ion exchange. Tests up to the end of the demonstration show that the CWF dissolves at a rate equal to or less than that for reference DHLW borosilicate glass. This suggests that the CWF repository performance will be comparable to that of the reference borosilicate glass. PCTs conducted on the CWF produced with salt from treating the 100

---

[39]ASTM C1174-98, "Standard Practice for Prediction of the Long Term Behavior of Materials, Including Waste Forms, Used in Engineered Barrier Systems (EBS) for Geological Disposal of High Level of Radioactive Waste," American Society for Testing and Materials, West Conshohocken, PA, 1998.

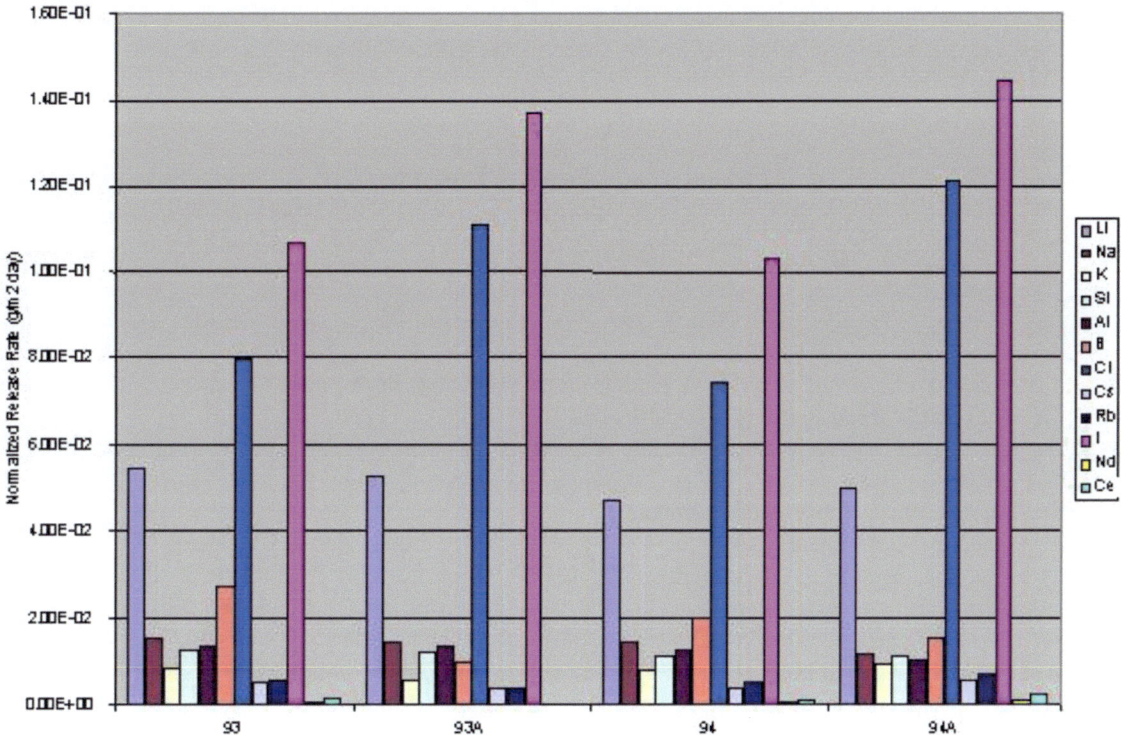

FIGURE 4.2 Product consistency test results for witness tubes and HIP cans (witness tubes results are denoted with the letter A). Witness tubes are small stainless steel samples used in HIP tests instead of the larger HIP cans as a means to get representative samples.
SOURCE: Argonne National Laboratory.

driver assemblies show that release of the rare earth elements, uranium, and plutonium are very low and are consistent. Other PCTs on irradiated CWF show anomalously high normalized rates of release for chloride ion and iodide ion compared to other elements (Figure 4.2). These results suggest that some fraction of chloride and iodide are not encapsulated within the sodalite or glass phases. With the exception of tests of radioactive samples and tests of greater than one year in duration, demonstration waste form testing matrices were completed by the end of the demonstration.

Finding: Dissolution tests on the CWF over a 6-month period indicate that the CWF dissolves at a rate equal to or less than that of reference Defense Program high level waste borosilicate glass.

Finding: If the long-term release of radionuclides from the CWF is found to control the dissolution of the inert borosilicate glass matrix, and if there is no change in the long-term dissolution rate of the glass matrix under repository conditions, the dissolution rate release performance of the CWF will be at least comparable to borosilicate glass.

Finding: The minor component actinides and rare earths form phases separate from sodalite and glass. The actinides occur as nano-size (colloidal) crystal inclusions associated with the glass or the glass/sodalite grain boundaries.

Finding: It is possible that some of these colloidal-sized crystal inclusions may be leached from the grain boundaries and that some may become colloidal suspensions with mobility much greater than expected from their solubility.[40]

Mechanical and Physical Properties

Several mechanical and physical properties of the CWF were determined by ANL: its cracking factor, thermal stability, fracture toughness, and density. The first two have a direct impact on CWF behavior in a geologic repository. The extent of cracking increases the amount of surface area exposed to the aqueous environment. Thermal stability testing of the CWF up to 500 °C provides a measure of the physical integrity, phase stability, and corrosion resistance under conditions that are more aggressive than those anticipated in a repository. Fracture toughness is an indicator of how the waste form surface area may change in response to unexpected impact events. The density measurements relate to product consistency.

No statistically significant variation in fracture toughness was observed by ANL among and within CWF samples. The fracture toughness of the CWF is comparable to measured values for borosilicate high-level waste glass from the Defense Waste Processing Facility.

Based on cracking measurement results, the CWF is expected to experience less cracking than DHLW glass in response to an impact. Heating of the reference CWF to 500 °C for at least 1 year shows no discernable change in physical characteristics or phase composition. The corrosion behavior is unchanged after 12 weeks of exposure to a 500 °C environment, with tests in progress on samples exposed for 1 year.

The density of CWF material produced under differing processing conditions shows no significant variation. The density profile measurements determined from different locations within a HIP similarly show no significant variation. This indicates that the HIP process is robust within the developed operating parameters specified for the EBR-II Spent Nuclear Fuel Treatment Demonstration Project. The density of the first radioactive CWF, formed from the electrorefiner salt from 100 driver fuel assemblies, agrees well with the nonradioactive, demonstration-scale CWF density, i.e., 2.35 versus 2.36 g/cm^3, respectively.

Finding: The mechanical and physical properties of the CWF are comparable to or better than those of DHLW borosilicate glass. Good product consistency is achieved using the specified demonstration HIP process parameters.

Finding: The physical and mechanical behavior of the CWF under repository conditions should be comparable to that of borosilicate high level waste glass.

Modeling

Performance waste form modeling has been carried out at ANL to predict the environmental impact of the ANL ceramic and metal waste forms on the operation of the proposed repository at Yucca Mountain. A simplified schematic of CWF degradation and radionuclide release is shown in Figure 4.3. The model is based on a transition state theory approach for the rate of silicate mineral dissolution.[41,42] The model incorporates two effects. The first of these is the forward reaction rate in the absence of dissolved silicic acid. The forward rate is both temperature

---

[40]For a study on the potential impact of actinides on repository performance, see A.B. Kersting, D.W. Efurd, D.L. Finnegan, D.J. Rokop, D.K. Smith, J.L. Thompson, "Migration of Plutonium in Ground Water at the Nevada Test," *Nature*, Vol. 397, 1999, pp. 56-59.

[41]A. Lasaga, "Transition State Theory," in *Reviews in Mineralogy Volume 8, Kinetics of Geochemical Processes*, A. Lasaga and R. Kirkpatrick, editors, Mineralogy Society of America, Washington, D.C., 1981.

[42]W. Bourcier, "Overview of Chemical Modeling of Nuclear Waste Glass Dissolution" in *Materials Research Society Symposium Proceedings*, Vol. 212, T. Abrajano and L. Johnson, editors, Materials Research Society, Pittsburgh, PA, 1991, pp. 3-18.

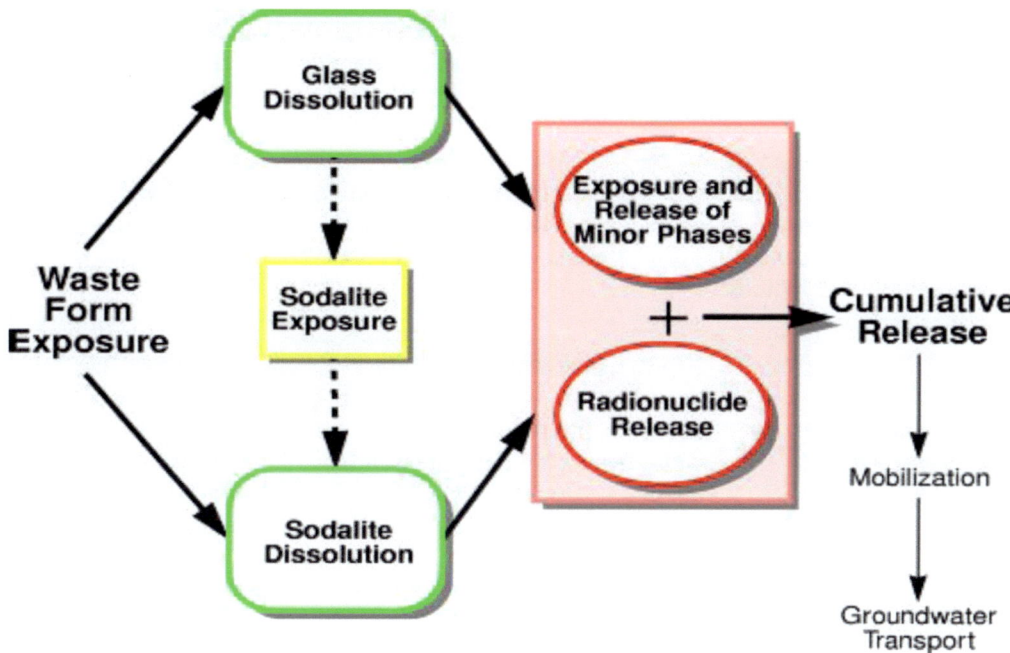

FIGURE 4.3 Ceramic waste form degradation and radionuclide release. This figure assumes glass is less durable than sodalite so that glass dissolution leads to additional sodalite exposure.
SOURCE: Argonne National Laboratory.

and pH dependent. The second effect takes into account the relative degree of saturation of silicic acid with respect to the solubility product of the dissolving solid. In previous applications to borosilicate glass, the solubility product of a proxy solid phase (e.g., amorphous silica) was used.

This approach has been used to model dissolution rates of high-level and low-level borosilicate waste glasses for performance assessments. ANL's use of the model is aimed at predicting degradation and radionuclide release for the CWF. The model developed for CWF attempts to predict long-term behavior based on short-term data. The final input data to the model will be based on MCC-1 (forward rate) and PCT data (relative saturation) on both simulated and ceramic waste forms.

The model assumes that the release of radionuclides from the CWF will be controlled by the dissolution of the major phases, sodalite and glass. The overall degradation rate of the CWF is assumed to be controlled by glass dissolution.[43] To date the model parameters being used are those of HLW glass. It is expected that HLW dissolution rates will bound those for the CWF. Parameters based on the CWF will be incorporated when complete experimental data on the pH and temperature dependence of dissolution of the CWF are available.

The model further assumes that the minor phases have uniform distribution and zero durability, that they are exposed through glass and sodalite dissolution, and that radionuclides in the minor phases are immediately released and available for transport once exposed. As noted previously, the radionuclide-bearing minor phases are known to be located along the grain boundaries of sodalite and glass and could be preferentially exposed over time.

---

[43]R.A. Wigeland, L.L. Briggs, T.H. Fanning, E.E. Feldman, E.E. Morris, and M.C. Petri, *Waste Form Degradation and Repository Performance Modeling*, NT Technical Memorandum No. 117, Argonne National Laboratory, Argonne, IL, 1999.

The model assumes sustained congruent dissolution of sodalite; i.e., aluminum and silicon are released to solution at a rate equal to their stoichiometric proportions in the sodalite phase. Sodalite is a high aluminum phase with Si/Al = 1. The model for the long-term dissolution rate of sodalite is based on the concentraton of silicic acid in solution. This assumption has been questioned by ANL and experiments are in progress to investigate the role of dissolved aluminum in sodalite dissolution.

A large part of the work in the future will be to refine and update the existing model (based on HLW glass dissolution) with the experimental data for rate of dissolution of the CWF as they become available. Additional features will also be added to the model such as effects of in-package chemistry changes, and the effect of colloidal particles as measured experimentally.

Finding: ANL's performance waste form model assumes that (1) the degradation of the CWF is controlled by dissolution of the major components, sodalite and glass, specifically by the rate of glass dissolution; (2) the minor phases have uniform distribution, and the radionuclides in the minor phases are exposed through glass and sodalite dissolution; (3) sodalite dissolves congruently, i.e., aluminum and silicon are released at the same rate; and (4) no credit is taken for the dissolution behavior of the minor phases.

Finding: The characteristics identified and tests conducted to date have shown that the third performance waste form model assumption listed above is not valid. The model must be refined with additional experimental test data before it can be used reliably to predict CWF behavior in a repository.

Finding: The dissolution rate model for the glass matrix of the CWF as developed to date suggests that this material should perform like borosilicate glass, and given the small quantity to be emplaced would have a negligible impact on the overall repository performance.

Finding: The success criteria (Chapter 6) regarding the CWF have been met, although it is recognized that further data collection and analysis must be carried out in the post-demonstration period to support a final decision on CWF acceptability for repository disposal.

Finding: The committee sees no significant barriers to successful demonstration of an acceptable CWF, although full testing will extend beyond the demonstration time frame.

RECOVERED URANIUM MATERIAL

As part of the EMT process, uranium metal is recovered at the cathode in the electrorefiner. The uranium metal is deposited as a dendritic growth that entraps as much as 20 wt % processing salt (with associated fission products and actinides). ANL's Technical Memorandum No. 106 (NT-106, figure 6) describes the remaining steps in the recovery and stabilization of this material.[44]

The entrapped chloride salt is distilled from molten uranium in a cathode processor operated at 1200 °C. A small quantity of the salt remains dissolved in the molten uranium pool and a small quantity of salt reacts with the crucible, resulting in fission product and actinide contamination of both. The presence and concentration of these impurities determine and limit the subsequent disposition options of the uranium material.

During the cathode processing step and/or the casting step, natural or depleted uranium (DU) is added to the highly enriched uranium (HEU) derived from the EBR-II driver fuel. The purpose of this downblending is to avoid proliferation issues with respect to subsequent disposition options. This step reduces the enrichment from

---

[44]R.W. Benedict, H.F. McFarlane, S.P. Henslee, M.J. Lineberry, D.P. Abraham, J.P. Ackerman, R.K. Ahluwalia, H.E. Garcia, E.C. Gay, K.M. Goff, S.G. Johnson, R.D. Mariani, S. McDeavitt, C. Pereira, P.D. Roach, S.R. Sherman, B.R. Westphal, R.A. Wigeland, and J.L. Willit, *Spent Fuel Demonstration Final Report*, ANL Technical Memorandum No. 106, Argonne National Laboratory, Argonne, IL, 1999, Figure 6.

~63 wt % ^{235}U to less than 50 wt %. It may be feasible to use the U recovered from EBR-II blanket fuel to accomplish this downblending.

The disposition options for this recovered uranium material are constrained by several DOE programmatic decisions and environmental impact statements.[45,46,47] ANL notes that "the depleted uranium byproduct does not have an identified potential commercial application at this time."[48] This is attributable to the fact that the radioisotopic composition of the recovered uranium does not meet the AST standard[49] for the fabrication of commercial nuclear fuel.

ANL further asserts that absent this commercial fuel fabrication option, "the [enriched, driver-derived] uranium could simply be stored indefinitely at low cost or disposed as low-level waste."[50] This latter option would require further downblending to 0.9 wt % ^{235}U.

A more speculative option is for the enriched uranium's transfer to DOE's Pu-disposition program.[51] The depleted uranium recovered from treatment of the EBR-II blanket fuel is currently limited to indefinite storage or disposal as a transuranic (TRU) classified waste.[52]

Finding: The current alternatives for disposition of uranium recovered from EBR-II fuel by electrorefining are limited to indefinite storage or speculative schemes for disposal.

Recommendation: The DOE should evaluate and select among these existing options for the disposition of recovered uranium in a timely manner so that the overall impacts of the EMT approach can be assessed.

---

[45]U.S. Department of Energy, *Disposition of Surplus Highly Enriched Uranium Final Environmental Impact Statement*, DOE/EIS-0240, June 1996.

[46]U.S. Department of Energy, *Final Programmatic Environmental Impact Statement for Alternative Strategies for the Long-Term Management and Use of Depleted Uranium Hexafluoride*, DOE/EIS-0269, April 1999.

[47]U.S. Department of Energy, *Storage and Disposition of Weapons-usable Fissile Materials Final Programmatic Environmental Impact Statement*, DOE/EIS-0229, December 1996.

[48]H.F. McFarlane, K.M. Goff, T.J. Battisti, B.R. Westphal, and R.D. Mariani, *Options for Disposition of Uranium Recovered from Electrometallurgical Treatment of Sodium-Bonded Spent Nuclear Fuel*, NT Technical Memorandum No. 109, Argonne National Laboratory, Argonne, IL, 1999.

[49]H.F. McFarlane, K.M. Goff, T.J. Battisti, B.R. Westphal, and R.D. Mariani, *Options for Disposition of Uranium Recovered from Electrometallurgical Treatment of Sodium-Bonded Spent Nuclear Fuel*, NT Technical Memorandum No. 109, Argonne National Laboratory, Argonne, IL, 1999, Reference 11.

[50]H.F. McFarlane, K.M. Goff, T.J. Battisti, B.R. Westphal, and R.D. Mariani, *Options for Disposition of Uranium Recovered from Electrometallurgical Treatment of Sodium-Bonded Spent Nuclear Fuel*, NT Technical Memorandum No. 109, Argonne National Laboratory, Argonne, IL, 1999.

[51]H.F. McFarlane, K.M. Goff, T.J. Battisti, B.R. Westphal, and R.D. Mariani, *Options for the Disposition of Uranium Recovered from Electrometallurgical Treatment of Sodium-Bonded Spent Nuclear Fuel*, NT Technical Memorandum No. 109, Argonne National Laboratory, Argonne, IL, 1999, p. 10.

[52]If DOE decides to dispose of the depleted uranium byproduct of sodium-bonded blanket fuel treatment, the material would probably have to be treated as transuranic waste rather than low-level waste. If disposition as transuranic waste were to be selected, there is no current impediment to shipping the material to New Mexico for disposal at WIPP.

5

Post-demonstration Activities

The DOE is preparing an environmental impact statement (EIS) for the treatment and management of the remaining sodium-bonded spent nuclear fuel in its inventory using the EMT process. Table 5.1 lists the DOE sodium-bonded fuels by category, quantity, characteristics, and storage location. The total sodium-bonded fuel in the DOE inventory is about 60 metric tons heavy metal (MTHM). The EIS preparation process began in February 1999 with the publishing of the Notice of Intent; a Record of Decision is expected in early 2000.

If the EMT process is to be used successfully to treat sodium-bonded fuels, or any other spent fuels,[1] a number of activities must be carried out after the EMT demonstration, which concluded in June 1999. First and foremost, ANL must complete all the activities required to qualify both the metal and ceramic waste baseline forms for repository disposal. Although not formally part of the demonstration, at the urging of the committee ANL developed test plans and began acquiring waste form performance data prior to the end of the demonstration.[2] The status of waste form development and qualification was reviewed in this committee's ninth report.[3]

In addition to efforts to address these waste form issues, at least two other post-demonstration efforts are essential if the remaining sodium-bonded fuel is to be treated successfully. First, ANL-E must provide ongoing technical support to operations at ANL-W, and ANL-W must complete the required facility modifications and qualify the new, larger-scale equipment needed to handle the increased volume of fuel. These constitute a minimum set of post-demonstration activities.

Other activities that were either abandoned or considerably reduced in scope, or were started too late during the demonstration project, could be considered for continued R&D in the post-demonstration period. These include the following:

- Determine the feasibility of pressureless sintering for producing a ceramic waste form and if feasible, qualify the waste form;
- Develop a high-throughput electrorefiner (HTER);

---

[1]Office of Civilian Radioactive Waste Management, *A Roadmap for Developing Accelerator Transmutation of Waste (ATW) Technology, A Report to Congress*, DOE/RW-0519, U.S. Department of Energy, Washington, D.C., 1999.

[2]National Research Council, *Electrometallurgical Techniques for DOE Spent Fuel Treatment: A Status Report on Argonne National Laboratory's R&D Activity*, National Academy Press, Washington, D.C., 1996, p. 2.

[3]National Research Council, *Electrometallurgical Techniques for DOE Spent Fuel Treatment: An Assessment of Waste Form Development and Characterization*, National Academy Press, Washington, D.C., 1999.

TABLE 5.1 DOE Sodium-bonded Spent Nuclear Fuel Inventory

Fuel Category	Quantity[a] (MTHM)	Characteristics	Storage Location
EBR-II driver	1.1	U metal 10% Zr alloy	ANL-W
EBR-II driver	2.0	U metal 5% fission alloy[b]	INTEC[c]
EBR-II blanket	22.4	DU[d] metal	ANL-W
Fermi-1 blanket	34.2	DU[d] metal Mo alloy	INTEC[c]
FFTF[e]	0.25	U metal Mo alloy	Hanford and ANL-W

[a]Pre-demonstration values.
[b]Fission alloy contains Mo, Ru, Rh, Pd, Zr, and Nb.
[c]Idaho Nuclear Technology and Engineering Center, located at Idaho National Engineering and Environmental Laboratory (INEEL).
[d]DU = depleted uranium.
[e]FFTF = Fast Flux Test Facility.

- Complete the development of the zeolite column to separate plutonium and fission products from the salt; and, finally,
- Complete the development of the lithium oxide reduction step as a front end-process to treat oxide fuel.

Of these post-demonstration activities, pressureless sintering and HTER and zeolite column development are required only if it could be shown that their implementation would significantly reduce the cost or the time required to treat the remaining sodium-bonded fuel. The remaining activity, development of the lithium oxide reduction step, is required if a decision is made by DOE to treat that fraction of the sodium-bonded EBR-II fuel, referred to as "disrupted" EBR-II fuel, for which the cladding has been breached and deterioration by oxide formation has occurred, or if DOE wishes to develop EMT for processing oxide fuels. Alternatively, the lithium oxide reduction work should continue if DOE wants an electrometallurgical process development that can treat oxide fuels.

In the remainder of this chapter, the committee discusses the post-demonstration activities planned by ANL and offers related recommendations.

WASTE FORM QUALIFICATION FROM A REPOSITORY PERSPECTIVE

The specific data required for waste form qualification are determined by the need to ensure the long-term safety of a deep geologic repository containing such waste forms. DOE-RW is preparing, but has not yet finalized, acceptance criteria for DOE spent nuclear fuel and high level waste.

The technical basis for such acceptance criteria has been addressed in previously published safety assessments for the proposed repository at Yucca Mountain, Nevada. The DOE-RW Yucca Mountain Project has conducted several system studies on repository safety. The total system performance assessment study in 1995 (TSPA-95),[4] in particular, reviews the technical basis for data needs with respect to waste-form and repository performance. The TSPA-95 report makes clear that there are several aspects to waste-form performance that assure safe levels for radionuclide releases from a repository.

The first aspect is the waste-matrix dissolution rate, also called the alteration or leach rate, that controls the long-term release of soluble radioactive elements. The second aspect is the solubility-limited concentration for a given radioactive element, imposed either by equilibrium between groundwater and a stable waste form matrix or by equilibrium between groundwater and alteration products that form because they are more stable than the dissolving waste form matrix. These aspects correspond exactly to those identified in a previous NRC study (the

---

[4]Office of Civilian Radioactive Waste Management, *Total System Performance Assessment Viability Assessment*, B00000000-01717-4301-00005, U.S. Department of Energy, Washington, D.C., 1995, Chapter 6.

Waste Isolation System Panel—WISP—report).[5] A third aspect is the potential formation of radionuclide-bearing colloids arising from waste form dissolution.

Each of these aspects will be influenced by environmental and design factors of the repository system. For example, dissolution rate will be a function of groundwater composition, temperature, and surface area contacted by groundwater over time. Solubility constraints, while also a function of groundwater composition and temperature, are intensive parameters that are not surface-area dependent.

Post-demonstration qualification testing of EMT-produced waste forms must therefore focus on the following:

- Long-term dissolution rate of the initial waste-form matrix (including the effects of fractures),
- Formation of radioactive element solubility-limiting solids, and
- Potential formation of radionuclide-bearing colloids.

These aspects of waste form qualification as they apply specifically to EMT waste forms are discussed in the following sections.

Long-term Dissolution (Corrosion) Rate of EMT Waste Forms

The waste acceptance product specifications (WAPS) include short-term product consistency tests (PCTs) to determine the initial dissolution rate (normalized leach rate) of the EMT waste forms. In Chapter 4, the committee expresses reservations about whether the current WAPS, based on WAPS developed for monolithic borosilicate glass, are appropriate for multiphase EMT waste forms. In particular, normalized leach rates of stable elements (e.g., Na, Si) that are common to two phases in the CWF, namely sodalite and borosilicate binder, are likely to be difficult to interpret unambiguously. More importantly, the radionuclide release behavior of other phases in the CWF that do not contain these stable components, such as U-Pu oxide, actinide silicate, and oxychlorides, cannot be determined by a conventional PCT devised for monolithic borosilicate glass.

Recommendation: In its post-demonstration activities, ANL should reevaluate the appropriateness and applicability of its overall model to address the dissolution behavior and the multiphase nature of the EMT waste forms, especially the CWF. Associated test protocols, including that for the current product consistency test (PCT), should also be reevaluated.

Formation of Radioelement Solubility-limiting Solids

A previous NRC report criticized "leach rate" as a measure of the long-term performance of waste forms under expected repository conditions.[6] The key concern is that short-term tests on waste forms under closed-system conditions fail to incorporate mass-transfer constraints (e.g., diffusive or diffusive-advective transport). Consideration of mass-transfer constraints shows that ultimately two factors control the long-term release of most radionuclides from a multibarrier repository. These are radioelement solubility limits, which are imposed either by the solubility of a stable waste matrix or formation of more stable alteration products, and fracturing, which can expose more surface area that contains the high-alpha-activity phases.[7,8]

---

[5]National Research Council, *A Study of the Isolation System for Geologic Disposal of Radioactive Waste*, National Academy Press, Washington, D.C., 1983.

[6]National Research Council, *A Study of the Isolation System for Geologic Disposal of Radioactive Waste*, National Academy Press, Washington, D.C., 1983.

[7]National Research Council, *A Study of the Isolation System for Geologic Disposal of Radioactive Waste*, National Academy Press, Washington, D.C., 1983.

[8]Nuclear Energy Agency, *The Status of Near-Field Modeling*, Organization for Economic Cooperation and Development, Paris, France, 1993.

Especially germane to the charge of this committee, however, is the recognition that solubility limits that determine long-term repository performance are strongly dependent on "the compositional parameters of the concentration-limiting solid phase."[9] Slight changes in chemical factors (e.g., waste-form composition, groundwater composition) can lead to different solubility-limiting solids with extreme differences (by several orders of magnitude) in predicted solubility concentrations for radioelements.[10]

The post-demonstration evaluation of the long-term performance of EMT waste forms, especially the CWF, under repository conditions must address this aspect of solubility limits for radioelements. Such an effort would address the formation and characterization of alteration phases under appropriate conditions, as well as measurement of the solubility-limited concentrations of radioelements. ANL cites the American Society for Testing and Materials (ASTM) Standard-C1174-97[11] as a basis for its test program for qualifying waste forms. With respect to predicting the long-term behavior of waste forms, the 1997 ASTM standard makes no reference to the perspectives and conclusions of the WISP report[12] regarding the formation of more stable, solubility-limiting solids. Nor does the ASTM standard reference the TSPA-95 safety assessment report[13] or equivalent system-level safety analyses that affirm the importance of solubility-limiting phases. A possible reason is that the ASTM standard restricts itself solely to the "alteration" of initial waste forms and other barrier materials.[14] As previously noted,[15] the alteration rate (leach rate) of waste forms is an irrelevant factor in the overall release rate of most radionuclides from waste forms emplaced in deep geologic repositories.

The committee notes that ANL has used a bulk dissolution rate (leach) test method developed for a single-phase nuclear waste form (borosilicate glass). It is now clear, however, that the CWF and the MWF are multiphase waste forms and that each phase has a very different radionuclide composition. Such different phases will likely experience different rates of dissolution under repository conditions. Hence, the present use of bulk leach rate tests has extremely limited value in assessing any meaningful measure of the performance of multiphase EMT waste forms in a geologic repository.

Recommendation: In the post-demonstration period, ANL should supplement and refine its current ASTM-based test protocols for waste form dissolution with respect to the technical perspectives on the long-term performance of the waste forms in geologic repositories, as described in the NRC's 1983 report by the Waste Isolation System Panel (WISP).[16]

---

[9]Office of Civilian Radioactive Waste Management, *Total System Performance Assessment Viability Assessment*, B00000000-01717-4301-00005, U.S. Department of Energy, Washington, D.C., 1995, p. 6-6.

[10]Office of Civilian Radioactive Waste Management, *Total System Performance Assessment for Viability Assessment*, B00000000-01717-4301-00005, U.S. Department of Energy, Washington, D.C., 1995.

[11]ASTM C1174-97, "Standard Practice for Prediction of the Long-term Behavior of Materials, Including Waste Forms, Used in Engineered Barrier Systems (EBS) for Geological Disposal of High-Level Radioactive Waste," American Society for Testing and Materials, West Conshohocken, PA, 1997.

[12]National Research Council, *A Study of the Isolation System for Geologic Disposal of Radioactive Waste*, National Academy Press, Washington, D.C., 1983.

[13]Office of Civilian Radioactive Waste Management, *Total System Performance Assessment Viability Assessment*, B00000000-01717-4301-00005, U.S. Department of Energy, Washington, D.C., 1995.

[14]ASTM C1174-97, "Standard Practice for Prediction of the Long-term Behavior of Materials, Including Waste Forms, Used in Engineered Barrier Systems (EBS) for Geological Disposal of High-Level Radioactive Waste," American Society for Testing and Materials, West Conshohocken, PA, 1997, Figure 1.

[15]National Research Council, *A Study of the Isolation System for Geologic Disposal of Radioactive Waste*, National Academy Press, Washington D.C., 1983.

[16]National Research Council, *A Study of the Isolation System for Geologic Disposal of Radioactive Wastes*, National Academy Press, Washington, D.C., 1983.

Potential Formation of Radionuclide-bearing Colloids

There is considerable concern regarding the potential for rapid migration of significant quantities of radionuclides, especially Pu, at Yucca Mountain via colloidal transport.[17] The as-produced specimens of the CWF contain separate U-Pu oxide particles on the order of 10 nm in radius. Such separate U-Pu oxide phases could conceivably contribute to colloidal transport upon waste form dissolution under repository conditions.

Finding: Analysis of the potential for formation and transport of radionuclide-bearing colloids should be specifically addressed in post-demonstration evaluation of EMT waste forms.

POTENTIAL FOR ALTERNATIVE, NONTESTING STRATEGIES FOR WASTE ACCEPTANCE

The committee notes that assessment of repository safety will be a function of the performance of all waste emplaced in such a geologic disposal system. The number of containers and the total DOE radionuclide inventory of EMT waste forms are extremely minor in comparison with the volume of commercial spent nuclear fuel and defense waste processing facility borosilicate glass intended for co-disposal at Yucca Mountain. It is conceivable that the uncertainties in radionuclide inventory and release-rate performance of these dominant waste forms may have a far greater impact on meeting a total-system safety standard than would conservative bounding assumptions made regarding EMT waste forms.

Alternatively, an arbitrary 10,000-year cutoff may be applied by the Environmental Protection Agency in its eventual safety standard for Yucca Mountain. In such a case, isolation strategies involving high-integrity containers designed to physically isolate HLW for more than 10,000 years could obviate the need for any long-term performance testing of waste forms.[18] As noted in this committee's fourth report, the 1983 WISP report argued that extrapolations from leach tests may not be "applicable to predicting performance of waste packages in geologic repositories."[19,20] Furthermore, even if a base case of extended containment were to be accepted, the potential for early failures in containers, arising from fabrication or emplacement operations, would still require a database on waste form performance.

Finding: There may be alternative, nontesting approaches to assessing the acceptability of EMT waste forms for geologic disposal and that the merits of these alternatives would have to be technically evaluated by the DOE and by other independent peer reviews.

Recommendation: The eventual DOE waste acceptance criteria for geologic disposal should take into account available technical assessments.[21] These waste acceptance criteria should be independently reviewed.

---

[17]Office of Civilian Radioactive Waste Management, *Viability Assessment of a Repository at Yucca Mountain*, DOE/RW-0508, U.S. Department of Energy, Washington, D.C., 1998.

[18]National Research Council, *Electrometallurgical Techniques for DOE Spent Fuel Treatment: Spring 1998 Status Report on Argonne National Laboratory's R&D Activity*, National Academy Press, Washington, D.C., 1998, pp. 13-14.

[19]National Research Council, *Electrometallurgical Techniques for DOE Spent Fuel Treatment: A Status Report on Argonne National Laboratory's R&D Activity*, National Academy Press, Washington, D.C., 1996, p. 8.

[20]National Research Council, *A Study of the Isolation System for Geologic Disposal of Radioactive Waste*, National Academy Press, Washington D.C., 1983, p. 6.

[21]See, for example, U.S. Nuclear Waste Technical Review Board, *Report to the U.S. Congress and the Secretary of Energy*, NWTRB, Arlington, VA, 1999, and references cited therein.

NEW PROCESS EQUIPMENT AND FACILITY MODIFICATIONS

Should DOE decide to use electrometallurgical technology to further process spent nuclear fuel in its inventory, most of the equipment used for the EMT demonstration process would be used as is during these post-demonstration operations. However, in addition to replacement of the driver-element chopper, the waste processing equipment will have to be resized to meet the increased throughput. For example, for the ceramic waste form a new V-mixer with double the capacity of the existing one, and a larger hot isostatic press (HIP) capable of processing a 51-cm-diameter can will be required. Additionally, a new furnace dedicated to producing the metal waste form is needed. This new furnace is intended to both distill the salt from the cladding and cast the cladding into an ingot. Design changes to the existing cathode processor are also necessary so that the crucible load capacity can be increased from 54 to 75 kg. During inventory operations, the cathode processor and casting furnace used in post- demonstration activities will be dedicated to uranium processing.

The facility will be modified to allow repair of equipment in the hot cell in such a way as to not disrupt ongoing operations. In addition to decreasing downtime due to equipment malfunctions, some parts of the infrastructure such as the cell purification and refrigeration systems will be upgraded.

Finding: Should DOE decide to treat the remaining sodium-bonded spent fuel inventory, continuing efforts will be required to increase the capacity of some process equipment and to modify facilities at ANL.

Recommendation: If the DOE decides to treat the remaining sodium-bonded spent fuel inventory and the waste form qualification efforts are successful, the required equipment upgrades and facility modifications should be adequately funded to ensure that treatment can be completed in a reasonable time and at a reasonable cost.

OTHER POSSIBLE ACTIVITIES

Pressureless Sintering as a Process for Preparing the Ceramic Waste Form

ANL is also investigating alternative methods to use of HIP for ceramic waste form fabrication. Such a method is "pressureless sintering."

The preparation steps for pressureless sintering are essentially identical to those for the HIP process. The process for pressureless sintering is discussed in detail in Chapter 3. Studies are being conducted to establish optimal fabrication procedures, and to establish whether pressureless sintering can produce a suitable waste form. The short-term leach characteristics of waste forms from HIP and pressureless sintering are also essentially identical, although the committee has previously noted that such short-term tests may not be fully indicative of long-term release-rate performance of waste forms under expected repository conditions. Furthermore, the heterogeneous nature of the multi-phase ceramic waste form mandates examination of the microstructure and phase composition of as-produced waste forms.

Finding: The use of pressureless sintering to produce the ceramic waste form can offer distinct advantages over the baseline HIP process. The potential advantages include a higher throughput per square foot of process space, increased safety, and reduced costs.

Recommendation: If pressureless sintering were to be used in place of the HIP process to produce the EMT ceramic waste form, waste form qualification studies would have to be conducted to determine its suitability for producing a waste form intended for deposit in a geologic repository.

Pressureless sintering may provide advantages relative to the HIP process during fabrication. The main technical advantage of the pressureless sintering over the HIP process would be its higher throughput per square foot of process space. It should be noted, however, that although the initial steps in the HIP process (i.e., loading, evacuating, and sealing of the HIP can) are carried out in the Ar cell, the actual HIP is carried out in air. Pressureless sintering would have to be carried out completely in the Ar cell. Also, the HIP process is inherently a batch process (load, bring to temperature and pressure and hold, and finally return to ambient temperature and pressure), whereas pressureless sintering, which depends only on maintaining a sample at a specified temperature for a specified length of time, can be operated in practice as a continuous process.

Continued Development of a High-throughput Electrorefiner (HTER)

In the post-demonstration period, a HTER, particularly if it could be cost-effectively developed and implemented in a timely fashion, could offer the advantage of considerably reducing the time required to treat the remaining sodium-bonded fuel. Currently, the proposed schedule calls for the remaining sodium-bonded spent fuel (approximately 58 metric tons of heavy metal-MTHM) to be treated at a rate of approximately 5 MTHM/year over a period of 12 to 13 years. The 5 MTHM/year rate is based on estimated rates of 0.6 MTHM/year and 4.4 MTHM/year for the Mark-IV and Mark-V, respectively. To achieve the estimated rate, the Mark-V will operate anode-cathode modules (ACMs) in three of the four ports simultaneously at a deposition rate of approximately 400 g/hour per ACM and a 50% duty cycle.

A processing rate of greater than 5 MTHM/year could potentially be achieved by further development of the 25-inch HTER at ANL-E. The committee notes that the rate of 5 MTHM/year rate could be doubled simply by adding a second Mark-V to the Ar cell at ANL-W. Since the development work on the Mark-V is essentially over and its capabilities have been demonstrated, this alternative appears to offer distinct cost advantages. Adding a second Mark-V will double the rate only if the electrorefining step is the rate-limiting step.

Finding: There are at least two options for increasing throughput up through the electrorefiner step in the EMT process. The first is continued development and implementation of a HTER (e.g., the 25-inch HTER under development at ANL-E) with a uranium deposition rate significantly exceeding that of the current Mark-V design. The second option is to simply double the current electrorefiner deposition rate by adding a second Mark-V electrorefiner to the Ar cell at ANL-W.

Recommendation: Continued development of a HTER should be evaluated in the context of the cost and time required for its development and implementation relative to the cost reduction that could be achieved by increasing the electrorefiner throughput by adding a second Mark-V and completing the inventory operations in the shorter time period.

Continued Development of the Zeolite Column

Initially the EMT process included both a multistage pyrocontactor and a zeolite column to treat (recycle) the spent electrorefiner salt. The purpose of the pyrocontactor was to remove the residual transuranic elements and the purpose of the zeolite column was to extract and immobilize the fission products. After this treatment, the salt was to be returned to the electrorefiner, and the waste-bearing zeolite from the column was to be converted to a solid monolithic form. This plan was later modified after the revised environmental assessment for the demonstration project prohibited using a cadmium cathode to separate out plutonium. As a result, the plutonium remained with the salt and the pyrocontactor became obsolete. However, work continued at ANL-E to develop the zeolite column. The plan was to process spent fuel so as to raise the level of fission product contamination to a point that as a result of running the salt through the zeolite column, the loading of the salt would be about 3 wt %. Because of this revised plan, radioactive waste treatment operations could not be started until late in the process, and characterization and qualification of a waste-load waste form would have to come even later. As a result, radioactive waste form production and characterization were not part of the original EMT demonstration project.

However, part way through the demonstration project, ANL changed directions and decided not to pursue salt recycle at that time. Instead, ANL implemented the so-called throw-away salt option, in which a portion of the plutonium- and fission-product-contaminated salt would be removed from the electrorefiner, and in a batch process, mixed directly with small zeolite and glass particles. This mixture would then be subjected to the HIP process to produce the desired waste form. The decision to implement the throw-away salt option resulted in essentially abandoning further radioactive zeolite column development. However at ANL-E, scale-up, remote design, and "cold" column operation development continued, although at a reduced level, in parallel with the EBR-II demonstration.

If the zeolite column work is to be resumed in the post-demonstration period, important questions remain such as how Pu and the other radioactive elements will load spatially on the column, the capacity of the columns, and the ability to maintain sufficiently uniform loading of Pu and fission products. Information is also needed on the effects of temperature and of potential temperature gradients on selectivity and sorption kinetics, as well as particle size and flow rate. Resolving the uncertainties related to these questions and issues would require a reasonable amount of R&D, but the problems should be ones that can be solved. However, the loss of water in the column in the early phases of elution may prove to be an intractable problem that would prevent the use of the column process. Small amounts of residual water on the zeolite are required to react with the metal oxychlorides to convert them to the desired oxide.[22]

Finding: The successful development of a production-scale zeolite column offers a number of significant technical challenges. The removal of water during the early stages of elution may prove to be an intractable problem that will prevent the successful development of a zeolite column compatible with the EMT process.

Finding: The volume of sodium-bonded spent fuel waste generated using the "throw away salt" option, where a portion of the plutonium and fission-product-contaminated salt is mixed directly with zeolite and glass particles for waste disposal, is such a small fraction of the total waste destined for geologic disposal that waste volume reduction resulting from the use of the zeolite column would not have a significant impact on the overall waste disposal problem.

Recommendation: Continued development of the zeolite column should not be considered a high priority unless a compelling argument can be made that its development and implementation would significantly reduce waste disposal costs or associated costs of EMT treatment of the DOE sodium-bonded spent fuel inventory.

Continued Development of the Lithium Reduction Front-end Process Step for Treating Oxide Fuels

For EMT to be used to treat oxide fuels, a head-end step is required to convert the oxide to metal. ANL-E has been pursuing the use of lithium metal as a reducing agent in molten LiCl salts to effect this conversion. The lithium metal is regenerated by electrolysis of the resulting lithium oxide so that the lithium and LiCl can be recycled. The interface between the reduction step and the electrorefining step is critical because any lithium or Li_2O carried over in the reduction could interact with UCl_3 in the electrorefiner. However, ANL has demonstrated the technical feasibility of coupling the reduction and electrorefining steps using existing technology.[23] However,

---

[22] National Research Council, *Electrometallurgical Techniques for DOE Spent Fuel Treatment: An Assessment of Waste Form Development and Characterization*, National Academy Press, Washington, D.C., 1999, p. 25.

[23] Argonne National Laboratory, *ANL Demonstration Project Monthly Highlights January 1998*, Argonne, IL, 1998.

some work remains to be done to complete the development of the Li reduction step. Some of the issues that remain are the extent of reduction of Pu_2O_3, the optimal design and materials of construction for the electrowinning cathode, the kinetics of UO_2 reduction, and methods for handling metallic lithium. Continued R&D is required, but the committee believes there are no "show stoppers."

Finding: The state of development of the lithium reduction head-end treatment step is fairly mature, and if it were allowed to go to completion, the DOE would have an additional option for treating uranium oxide spent nuclear fuel.

Recommendation: If the DOE wants an additional option besides PUREX for treating uranium oxide spent nuclear fuel, it should seriously consider continued development and implementation of the lithium reduction step as a head-end process to EMT.

6

Electrometallurgical Technology Demonstration Project Success Criteria

INTRODUCTION

The EBR-II Spent Nuclear Fuel Treatment Demonstration Project began in June 1996 and ended in June 1999. During the testing program, 100 EBR-II driver (core) fuel assemblies and up to 25 EBR-II blanket assemblies were to be processed. Waste form samples were to be prepared and tested. Sufficient data for evaluating the safety, environmental impact, and economics of the technology were to be generated.

The committee, throughout the three phases of its study, recommended that ANL adopt criteria to evaluate the success of its demonstration project.[1] The committee first proposed such criteria in its second report, recommending four separate criteria as accomplishments against which to evaluate the success of the demonstration project:[2]

- Demonstration of batch operation of an electrorefiner and a cathode processor of approximately 200 kg/day of radioactive EBR-II spent fuel without failure for about 30 days.
- Quantification (for both composition and mass) of recycle, waste, and product streams that demonstrate projected material balance with no significant deviations.
- Demonstration of an overall dependable and predictable process, considering uptime, repair and maintenance, and operability of linked process steps.
- Demonstration that releases of radioactivity remain at or below those levels anticipated and specified in equipment design and operating plans. Exposure of operating personnel to radiation must be minimal and must in all cases remain below limits set by the U.S. Nuclear Regulatory Commission.

In phase two of its study, the committee recommended the development of success criteria for the EBR-II spent fuel demonstration project: "A well-defined set of performance criteria needs to be developed. The criteria would provide ANL with a clear set of objectives. The achievement of those objectives would better position ANL to request approval to proceed to additional applications of its electrometallurgical technology program."[3] This

---

[1] See Appendix D, in which are reprinted the recommendations made in the committee's nine reports.
[2] National Research Council, *An Assessment of Continued R&D into an Electrometallurgical Approach for Treating DOE Spent Nuclear Fuel*, National Academy Press, Washington, D.C., 1995, pp. 5-7 and 5-8.
[3] National Research Council, *Electrometallurgical Techniques for DOE Spent Fuel Treatment: Fall 1996 Status Report on Argonne National Laboratory's R&D Activity*, National Academy Press, Washington, D.C., 1996, p. 7.

point was reiterated in the committee's sixth report: "Before the demonstration is completed, DOE should establish criteria for success in the demonstration phase to allow evaluation of the electrometallurgical technology for further use."[4] The criteria proposed by the committee were based on what the committee believed were reasonable estimates of a pilot-plant-scale production.

In preparation for operations related to the spent fuel demonstration project, ANL prepared an Operation Readiness Review for the fuel conditioning facility (FCF) where radioactive operations were to take place.[5] In August 1995, the DOE Material Control and Accountability Audit was completed.[6]

During the summer of 1995, several nongovernmental organizations had questioned the adequacy of the environmental assessment (EA) prepared under the IFR program. The resolution of environmental issues was all that was needed before operations with irradiated fuel could begin. During the fall of 1995, ANL decided to prepare a new environmental impact assessment to specifically address the spent fuel treatment operations in the FCF. This document was issued by DOE during late January 1996,[7] and the environmental evaluation was completed in the spring of 1996.

As a result of the new EA, the demonstration project was scaled back from initial estimates. The goal was to process 100 driver and 25 blanket fuel assemblies from EBR-II to demonstrate the electrometallurgical technique.[8]

In 1998, ANL proposed a set of performance criteria for the demonstration project, with goals to meet each criterion. The committee in its seventh report evaluated these criteria and goals and found them to be adequate for assessing the success of the demonstration project.[9]

DEMONSTRATION PROJECT SUCCESS CRITERIA

The criteria proposed by ANL in 1998 for the demonstration project were similar in scope to those recommended by the committee in 1995, but smaller in scale in order to conform to the EA. The four criteria address the process, the waste streams, and the safety of the electrometallurgical demonstration project. The four criteria and lists of goals are reprinted, with the exception noted, from pages 24 and 25 of the committee's seventh report.

Criterion 1: Demonstration that 100 driver and up to 25 blanket experimental breeder reactor assemblies can be treated in a fuel-conditioning facility (FCF) within 3 years, with a throughput rate of 16 kg per month for driver assemblies sustained for a minimum of 3 months and a blanket throughput rate of 150 kg per month sustained for 1 month.[10]

---

[4]National Research Council, *Electrometallurgical Techniques for DOE Spent Fuel Treatment: Status Report on Argonne National Laboratory's R&D Activity Through Spring 1997*, National Academy Press, Washington, D.C., 1997, p. 11.

[5]K.M. Goff, R.D. Mariani, D. Vaden, N.L. Bonomo, and S.S. Cunningham, "Fuel Conditioning Facility Electrorefiner Start-up Results," in *Proceeding of the Embedded Topical Meeting on DOE Spent Nuclear Fuel & Fissile Material Management, Reno, NV, June 16-20, 1996*, American Nuclear Society, Inc., La Grange Park, IL, pp. 137-143.

[6]Environment, Safety, and Health, *Material Control and Accountability Manual*, Lawrence Livermore National Laboratory, Livermore, CA, 1995.

[7]U. S. Department of Energy, *Draft Environmental Assessment – Electrometallurgical Treatment Research Demonstration Project in the Fuel Conditioning Facility at Argonne National Laboratory-West*, DOE/EA-1148, Washington, D.C., 1996.

[8]In June 1999, Robert G. Lange, associate director for Nuclear Facilities Management, Office of Nuclear Energy, Science and Technology, sent a letter to the National Research Council, clarifying the amount of EBR-II spent fuel that was treated in the demonstration project. In this letter Lange states that "The Finding of No Significant Impact issued in May 1996 for the environmental assessment of the demonstration project was based on the treatment of no more than 125 assemblies of EBR-II spent fuel (100 driver assemblies and 25 blanket assemblies)."

[9]R.W. Benedict, H.F. McFarlane, J.P. Ackerman, R.K. Ahluwalia, L.L. Briggs, H. Garcia, E.C. Gay, K.M. Goff, S.G. Johnson, R.D. Mariani, S. McDeavitt, G.A. McLennan, C. Pereira, P.D. Roach, and B.R. Westphal, *Spent Fuel Treatment Demonstration Interim Status Report*, NT Technical Memorandum No. 74, Argonne National Laboratory, Argonne, IL, 1998, Appendix A.

[10]The original statement of this criterion was as follows: "Demonstration that 125 EBR-II assemblies can be treated in a fuel-conditioning facility (FCF) within 3 years with a throughput rate of 16 kg/month for driver assemblies sustained for a minimum of 3 months and a blanket throughput rate of 150 kg per month sustained for 1 month." The intent of this success criterion was to define minimum required throughput rates for the treatment process to be met within 3 years using no more than the amount of EBR-II fuel that had been set aside. All 100 drivers

Specific goals to meet criterion 1:

1. Freeze process modifications and operating parameters while demonstrating a continuous throughput of 16 kg of driver uranium for 3 consecutive months.
2. Demonstrate the capability to electrorefine approximately 150 kg of blanket spent fuel in 1 month.
3. Distill the electrolyte from ER cathode products through the cathode processor in an FCF and blend the resulting ingot with depleted uranium in the casting furnace to produce a low-enriched uranium storage ingot.
4. Specify acceptable operating parameters and throughput for the cathode processor to meet uranium product specifications and ER production rates of 16 kg of driver uranium for 3 consecutive months.
5. Specify acceptable casting-furnace operating parameters for producing low-enriched uranium from 16-kg driver uranium per month for 3 consecutive months.
6. Cast three batches of irradiated cladding hulls (two driver assemblies per batch) into a typical metal waste form (stainless steel with 15 percent zirconium).
7. Process 3 kilograms of salt containing approximately 6 wt % fission products into 10 ceramic waste samples.

Criterion 2: Quantification (for both composition and mass) of recycle, waste, and product streams that demonstrate projected material balance with no significant deviations.

Specific goals to meet criterion 2:

1. Develop uranium product specifications with range of acceptable impurities: plutonium, neptunium, technetium-99 and ruthenium-106. Specify process-operating parameters for uranium ingots that meet uranium specifications.
2. Develop metal waste specifications that are based on performance characterization results of small samples with variations in the principal constituents: zirconium, uranium, technetium, plutonium, neptunium, and noble metals. Determine performance characterization with electrochemical technique, corrosion tests, vapor hydration tests, and attribute tests.
3. Develop metal waste process specifications for major process variables: operating temperatures, hold time, and cooling rate.
4. Develop ceramic waste specifications that are based on performance characterization results of samples with principal constituent variations: glass, fission products, uranium, and plutonium. Determine performance characteristics with attribute, characterization, accelerated, and service condition tests.
5. Develop ceramic waste process specifications for major process variables: free chloride, zeolite moisture content, and chloride per unit cell.
6. Quantify volume of low-level and transuranic waste generation under standard operating conditions.
7. Return the cathode processor condensate to the individual ERs during the 16-kg driver per month for 3 months and 150 kg blanket per month operations.
8. Specify unit process operations for metal spent fuel treatment, uranium ingot production, and waste form production.
9. Estimate mass balances for uranium, transuranics, sodium, and key fission products for overall process.
10. Prepare the flowsheet and develop process specifications for the subsequent inventory operation.

---

were treated to produce the needed throughput data and to obtain the fission product buildup required to test ceramic waste form production. However, it was projected that throughput and performance goals under this criterion for the blanket fuel would be met before 25 blanket assemblies would have been treated. Therefore, to ensure that this intent was properly understood, DOE recommended the change in wording for success criterion 1. The committee agreed with this change in wording.

Criterion 3: Demonstration of an overall dependable and predictable process, considering uptime, repair and maintenance, and operability of linked process steps.

Specific goals to meet criterion 3:

1. Record facility and equipment availability for process operations during the 3-month 16 kg per month driver demonstration.
2. Record process interruption for chemistry results during the 3-month operation at 16 kg per month.
3. Develop quantitative process models for each key step in the treatment process.
4. Develop a process model that estimates throughputs as a function of equipment availability, maintenance requirements, and individual process times.

Criterion 4: Demonstration that safety risks, environmental impacts, and nuclear materials accountancy are quantified and acceptable within regulatory limits.

Specific goals to meet criterion 4:

1. Demonstrate that the FCF air emissions result in an effective dose equivalent to the public less than 10 mrem per year, which is the limit in DOE 5400.5 and is less than the 25 mrem per year limit in the State of Idaho Permit to Construct Air Pollution Emitting Source.
2. Show that FCF personnel exposure is less than 0.5 rem per year average and 1.5 rem per year for the maximum individual exposure, which is a factor of 3 less than the DOE Occupational Radiation Protection Final Rule 10CFR835 limit that is 5 rem per year.
3. Demonstrate a material control and accountability system that shows the historical inventory difference for uranium and plutonium is within control limits based on variance propagation of measurement and sampling errors, as specified in DOE Order 5633.3B.
4. Record any unlikely and extremely unlikely accident (as defined in the Final Safety Analysis Report) during the demonstration.
5. Estimate the safety risks, environmental impacts, and material accountancy for the inventory operations.

COMMITTEE EVALUATION OF ANL'S DEMONSTRATION PROJECT BASED ON THE DEMONSTRATION PROJECT SUCCESS CRITERIA

The committee's evaluation of ANL's ability to meet the demonstration project success criteria derived from a number of factors. The committee received technical memoranda from ANL covering all aspects of its demonstration project. The committee also made yearly visits to both ANL-W, where the demonstration project took place, and ANL-E, where R&D was performed in support of the demonstration project. The committee also met with ANL personnel involved in the demonstration project in formal briefings at least three times each year. All of these sources of information were utilized in the committee's evaluation.

After the demonstration project was completed, the committee received a final overview of the technical accomplishments of the electrometallurgical demonstration project from Robert Benedict (ANL) in September 1999.[11] This briefing included the release to the committee of ANL final reports on the demonstration project. Following the September 1999 briefing, the committee reviewed each of the success criteria, together with each related goal to meet the success criteria, in order to come to an overall evaluation of the electrometallurgical demonstration project. The committee came to the following conclusions:

---

[11] See summary in Appendix C.

Criterion 1

Goal 1: The committee believes that this goal has not only been met, but the rate of processing has been exceeded.
Goal 2: The committee believes that this goal has been met.
Goal 3: The committee believes that this goal has been met.
Goal 4: The committee believes that this goal has been met.[12]
Goal 5: The committee believes that this goal has been met.
Goal 6: The committee believes that this goal has been met noting that the actual content of the as-produced ingots fits the target range of 5 to 20 percent Zr.
Goal 7: The committee believes that this goal has been met.

Criterion 2

Goal 1: The committee believes that this goal has been met.
Goal 2: The committee believes that this goal has been met. The committee notes that these are not results, they are methods.
Goal 3: The committee believes that this goal has been met.
Goal 4: The committee believes that this goal has been met. The committee notes that testing time has been limited. This statement is made not as a criticism, but as an observation. Although the committee believes that this goal has been met, this conclusion does not imply that either the metal waste form or the ceramic waste form are qualified waste forms. In addition, the committee recognizes that further data collection and analysis must be carried out in the post-demonstration period to support a final decision on MWF and CWF acceptability for repository disposal.
Goal 5: The committee believes that this goal has been met. For movement to larger HIP cans, further R&D is needed in the post-demonstration period. The process referred to in the goal is for small HIP cans.
Goal 6: The committee believes that this goal has been met.
Goal 7: The committee believes that this goal has been met.
Goal 8: The committee believes that this goal has been met.
Goal 9: The committee believes that this goal has been met. The committee notes that the data for the driver fuel is more important than the data for the blanket fuel.
Goal 10: The committee believes that this goal has been met.

Criterion 3

The committee notes that the demonstration was limited to a relatively short period at the end of the demonstration project period, as much of the early period of the overall demonstration project was spent on R&D.

Goal 1: The committee believes that this goal has been met.
Goal 2: The committee believes that this goal has been met.
Goal 3: The committee believes that this goal has been met.
Goal 4: The committee believes that this goal has been met.

Criterion 4

Goal 1: The committee believes that this goal has been met.
Goal 2: The committee believes that this goal has been met.

---

[12] A full description of the results obtained to meet this goal may be found in H.F. McFarlane, K.M. Goff, T.J. Battisti, B.R. Westphal, and R.D. Mariani, *Options for Disposition of Uranium Recovered from Electrometallurgical Treatment of Sodium-Bonded Spent Nuclear Fuel*, NT Technical Memorandum No. 109, Argonne National Laboratory, Argonne, IL, 1999.

Goal 3: The committee believes that this goal has been met.
Goal 4: The committee believes that this goal has been met.
Goal 5: The committee believes that this goal has been met, based on ANL's safety analysis.[13] Concerns with scale-up of the HIP process have been noted by the committee in previous reports.[14]

Findings and Summary

Finding: The committee finds that ANL has met all of the criteria developed for judging the success of its electrometallurgical demonstration project.

Finding: The committee finds no technical barriers to the use of electrometallurgical technology to process the remainder of the EBR-II fuel.

The EBR-II demonstration project has shown that the electrometallurgical technique can be used to treat sodium-bonded spent fuel. If the DOE decides to complete the treatment of EBR-II spent fuel and blanket material, the committee has found that there are no technical barriers to the use of EMT to achieve this objective. The major hurdle that remains is qualification of the waste forms from this processing. The total quantity is relatively small, particularly in comparison to the total DOE spent fuel inventory, so even if qualification of the waste form were to prove impossible, the quantity of these materials that had been produced would be modest. The committee has found no significant technical barriers in the use of electrometallurgical technology to treat EBR-II spent fuel, and EMT therefore represents a potentially viable technology for DOE spent nuclear fuel treatment. However, before using EMT for processing other spent fuels in the DOE inventory that would generate much larger amounts of these wastes than were produced in ANL's demonstration project, it would be necessary for these waste forms to receive the acceptance qualification.

---

[13]H.E. Garcia, C.H. Adams, D.B. Barber, R.G. Bucher, I. Charak, R.J. Forrester, S.J. Grammel, R.P. Grant, R.J. Page, D.Y. Pan, A.M. Yacout, L.L. Burke, and K.M. Goff, *Analysis of Spent Fuel Treatment Demonstration Operations*, NT Technical Memorandum No. 108, Argonne National Laboratory, Argonne, IL, 1999.

[14]National Research Council, *Electrometallurgical Techniques for DOE Spent Fuel Treatment: Status Report on Argonne National Laboratory's R&D Activity as of Fall 1998*, National Academy Press, Washington, D.C., 1999, p. 19.

Appendixes

APPENDIX A

Committee Charge and Statements of Task

The Committee on Electrometallurgical Techniques for DOE Spent Fuel Treatment has operated in three distinct phases. Each phase has had a charge and a statement of task that reflected the emphasis of the committee at that time.

The committee's charge and statement of task for phase three were clarified in a series of letters to the NRC from DOE. These letters are included in this appendix on pages 81 to 84.

COMMITTEE CHARGE—PHASE ONE

1. Hold a first meeting to receive briefings from representatives of the DOE and ANL, additional experts identified by the committee, and representatives of other relevant activities of the National Research Council and National Academies of Sciences and Engineering, and then prepare an interim report to address the question, "Do electrometallurgical techniques represent a potentially viable technology for DOE spent fuel treatment that warrants further research and development?"

2. Study in more depth the advantages and disadvantages of continued R&D into electrometallurgical processing as a candidate technology for disposition of DOE spent nuclear fuel, specifically addressing the issues of technical feasibility, cost-effectiveness, suitability of the metallic waste form for long-term storage or geologic disposal, and nonproliferation implications, and write a report on the committee's assessments.

Statement of Task—Phase One

The Board on Chemical Sciences and Technology (BCST) will organize a committee of approximately 12 scientific and technological experts to carry out a fast-track study of pyrometallurgical techniques as potential technologies for DOE spent fuel treatment. BCST will also utilize the results of CISAC's Reactor Panel report on disposition options for weapons plutonium. The BCST is well-suited to the proposed task, having institutional responsibility for the broad range of chemical issues including both chemistry and chemical engineering. The BCST previously organized the Separations Subpanel for the NRC Study on *Separations Technologies and Transmutation Systems*; that Subpanel conducted an evaluation of pyrometallurgical processing technologies from the perspective of spent fuel recycling.

This project has been discussed with both CISAC and BRWM and will be carried out with their assistance. CISAC and BRWM staff are included in the budget and will be involved in all phases of the project, beginning with identification of potential committee members with the necessary range of expertise.

Early in the study, an interim report will address the question, do pyrometallurgical techniques represent a potentially viable technology for DOE spent fuel treatment that warrants further research and development? After completing its study, the committee will write a final report assessing the advantages and disadvantages of continued research and development of pyrometallurgical processing as a candidate technology for disposition of spent fuel. Specific issues to be addressed in the study include technical feasibility; cost-effectiveness; suitability of the metallic waste form for long-term storage or geologic disposal; and nonproliferation implications.

COMMITTEE CHARGE—PHASE TWO

1. Carry out an ongoing evaluation of Argonne National Laboratory's (ANL's) research and development (R&D) activity on electrometallurgical techniques for treatment of DOE spent fuel, including their specific application to Experimental Breeder Reactor II (EBR-II) spent fuel.

2. Evaluate the scientific and technological issues associated with extending this R&D program to handle plutonium should the DOE decide that an electrometallurgical treatment option for the disposition of excess weapons plutonium is worth pursuing.

Statement of Task—Phase Two

Under the oversight of the Board on Chemical Sciences and Technology (BCST) the Committee on Electrometallurgical Techniques for DOE Spent Fuel Treatment will carry out two tasks. The first will be an ongoing evaluation of Argonne National Laboratorys R&D activity on electrometallurgical techniques for DOE spent fuel treatment, including their specific application to EBR-II spent fuel. The second task would be to evaluate the scientific and technological issues associated with extending this research and development program to handle plutonium, should the DOE decide that an electrometallurgical treatment option for the disposition of excess weapons plutonium is worth pursuing.

COMMITTEE CHARGE—PHASE THREE

1. Monitor the scientific and technical progress of the ANL program on electrometallurgical technology for the treatment of DOE spent nuclear fuel.

2. Examine the viability of electrometallurgical treatment technology in light of technical progress in other possible treatment technologies.

3. Evaluate the criteria by which the success of the demonstration project will be judged.

Department of Energy
Washington, DC 20585

October 31, 1997

Dr. Douglas J. Raber
National Academy of Sciences
National Research Council
2101 Constitution Avenue, N. W.
Washington, D.C. 20418

Dear Dr. Raber:

The Department is very pleased with the assessments on the Argonne National Laboratory electrometallurgical research and development activity carried out by the National Research Council, *Committee on Electrometallurgical Techniques for DOE Spent Fuel Treatment*. The Committee's reports have been of considerable value to the Department in evaluating the technical progress of the electrometallurgical research program.

The Committee's work thus far has covered three important areas:

- evaluating the viability of electrometallurgical technology for the treatment of DOE spent nuclear fuel and considering whether the technology warrants further research and development;

- conducting an ongoing evaluation of Argonne National Laboratory's R&D activity on electrometallurgical techniques for DOE spent fuel treatment, including their specific application to EBR-II spent fuel; and

- evaluating the scientific and technological issues associated with extending this research and development program to handle plutonium, should the DOE decide that an electrometallurgical treatment option for the disposition of excess weapons plutonium is worth pursuing.

As the Committee is aware, a variety of factors have combined to extend the schedule for the EBR-II spent fuel demonstration, which is now expected to be completed in June 1999. As a result, we request that your Committee continue its ongoing evaluation for the duration of this project. The Department requests that the Committee plan to issue a final report with final recommendations and observations subsequent to the completion of the demonstration project.

In addition, there may be other technologies available that might be relevant for treating some of the Department's spent nuclear fuel. In response to these developments, we request that the Committee, as part of its ongoing evaluation, review the viability of electrometallurgical treatment technology in light of technical progress in other possible treatment technologies. It is the Department's expectation that the Committee will assess the applicability of alternative technologies in considering the viability of electrometallurgical technology. Further, we request the Committee's evaluation of criteria developed by Argonne National Laboratory and the Department to determine the success of the demonstration project.

We request that the Committee's consideration of alternative treatment technologies and of the success criteria be provided to the Department in the next of its reports to be completed by April 1998.

Finally, we would like to recognize the excellent support provided to this activity by Dr. Tamae Wong over the last year. Dr. Wong's efforts have been appreciated by the Department and we will miss her as she moves on to her next assignment with the Council.

We appreciate the Council's support in continuing the evaluation of the electrometallurgical treatment research program. Please call me on (202) 586-6630 should you have any questions.

Sincerely,

William D. Magwood, IV
Associate Director for Planning and
  Analysis
Office of Nuclear Energy, Science
  and Technology

Department of Energy
Germantown, MD 20874-1290

February 12, 1998

Dr. Douglas J. Raber
National Academy of Sciences
National Research Council
2101 Constitution Avenue, N.W.
Washington, D.C. 20418

Dear Dr. Raber:

We are pleased that the National Research Council, "Committee on Electrometallurgical Techniques for DOE Spent Fuel Treatment," is continuing its assessments of Argonne National Laboratory's electrometallurgical research and development activities, pursuant to our letter of October 31, 1997. We appreciate the opportunity to coordinate with your staff in planning the committee's future activities.

To confirm these recent agreements between our staffs, we request that the committee include an assessment of the applicability of a "no-treatment" option in its review of alternatives to the electrometallurgical technology for the disposition of Experimental Breeder Reactor-II (EBR-II) spent nuclear fuel. Such an option could involve use of a canister and overpack or some other technology to allow emplacement of the EBR-II spent fuel in a geologic repository without treatment of the fuel. We will identify experts to present this disposition option to the committee at the March 16-17, 1998, sessions to address alternative technologies.

We look forward to the committee's consideration of alternative treatment technologies to be provided to the Department in the next committee report. Please call me at (301) 903-2915 should you have any questions.

Sincerely,

Robert G. Lange, Associate Director
  for Facilities
Office of Nuclear Energy,
  Science and Technology

Department of Energy
Washington, DC 20585

August 19, 1998

Dr. Douglas J. Raber
National Academy of Sciences
National Research Council
2101 Constitution Avenue, N.W.
Washington, D.C. 20418

Dear Dr. Raber:

As you requested, we are pleased to clarify the intent of the current task under which the *Committee on Electrometallurgical Techniques for DOE Spent Fuel Treatment* is reviewing alternatives to electrometallurgical treatment. In a letter to you from Mr. William Magwood dated October 31, 1997, we asked the committee to "review the viability of electrometallurgical treatment technology in light of technical progress in other possible treatment technologies." We intended that this review be undertaken specifically within the context of Experimental Breeder Reactor (EBR)-II sodium-bonded spent fuel since we are responsible for its dispositioning and are pursuing activities to support a Department of Energy decision on its treatment. This intent was also reflected in my letter to you dated February 12, 1998, which asked that the committee "include an assessment of the applicability of a 'no-treatment' option in its review of alternatives to the electrometallurgical technology for the disposition of EBR-II spent nuclear fuel."

We appreciate the council's support in continuing the evaluation of electrometallurgical research and hope this information is beneficial to your efforts to expeditiously publish the results of this review. Please call me at (301) 903-2915 should you have any questions.

Sincerely,

Robert G. Lange, Associate Director
 for Nuclear Facilities Management
Office of Nuclear Energy,
Science and Technology

APPENDIX B

Meeting Summary

**Meeting of the Committee on Electrometallurgical Techniques for DOE
Spent Fuel Treatment
Argonne National Laboratory–West
Idaho National Engineering and Environmental Laboratory
July 21-22, 1999**

JULY 21, 1999—OPEN SESSION—AGENDA

8:15 a.m.	Demonstration Status and Plans	R.W. Benedict
9:00 a.m.	Driver Processing Results	R.D. Mariani
9:30 a.m.	Cathode Processing and Casting Results	B.R. Westphal
10:00 a.m.	Break	
10:15 a.m.	Electrorefiner Throughput Studies	J.L. Willit
10:45 a.m.	Blanket Processing Results	S.R. Sherman
11:15 a.m.	Uranium Disposition Options	H.F. McFarlane
11:45 a.m.	Lunch	
12:30 p.m.	Ceramic Waste Process and Materials Studies	S. McDeavitt
1:00 p.m.	Ceramic Waste Demonstration Processing Results	K.M. Goff
1:30 p.m.	Ceramic Waste Qualification Testing	L.R. Morss
2:00 p.m.	Ceramic Waste Uranium/Plutonium Behavior Studies	S.G. Johnson/L.R. Morss
2:30 p.m.	Ceramic Waste Product Consistency Testing	T.P. O'Holleran
3:00 p.m.	Break	
3:15 p.m.	Metal Waste Qualification Testing	D. Abraham
3:45 p.m.	Metal Waste Product Testing	D.D. Keiser
4:15 p.m.	Metal Waste Release Modeling	M.C. Petri
4:45 p.m.	Repository Performance Modeling	E.E. Morris
5:15 p.m.	Adjourn	

SUMMARY OF PRESENTATIONS

Gregory Choppin, committee chair, opened the open session with an introduction of the committee members.

Robert W. Benedict, ANL, spoke on the EBR-II spent fuel demonstration status and plans. The EBR-II spent fuel treatment flow sheet was reviewed, demonstrating the separation of uranium, and the ceramic and metal waste forms individually. An overview of repeatability results for electrorefining was given. The specific success criterion relating to this issues states, "Freeze process modifications and operating parameters while demonstrating a continuous throughput of 16 kg of driver uranium for three months." The electrorefiner repeatability demonstration began on November 14, 1998, and ended on January 22, 1999, for a total of 61 working days. The average treatment rate was approximately 24 kg per month.

The Mark-V electrorefiner is used to treat blanket fuel. The success criterion relating to blanket fuel treatment requires a throughput rate of 150 kg per month sustained for 1 month. At the time of the meeting (July 21, 1999), consecutive treatment had started with three ports on the electrorefiner running in parallel, with hopes of getting the fourth port running. Significant achievements include the following: the latest run conditions allow 190 to 240 g of uranium per hour as an average production rate, four ports are operational, five blanket assemblies have been treated, and control software allows unattended operation.

Operations with irradiated fuel include both the cathode processor and the casting furnace. The cathode processor has treated 40 driver batches, 6 blanket batches, and 8 cladding hull batches. The casting furnace has treated 40 driver batches, 6 blanket batches, and 7 metal waste batches.

Significant accomplishments in the treatment process include the following: driver treatment has processed 100 driver assemblies, 8 assemblies were treated in 1 month, 1110 kg of low-enriched uranium were cast, the cathode processor batch size increased from 12 to 19 kg, and the casting furnace batch size increased from 36 to 54 kg. Blanket treatment has processed 5 of 25 blanket assemblies, the Mark-V electrorefiner has run 11 batches of irradiated blankets, the cathode processor has consolidated up to 42 kg of product in a batch, 125 kg of blanket product have been cast, and the blanket element chopper is operational.

Significant accomplishments in waste activities include the following. The stainless steel-zirconium alloy continues as the metal waste form. The test matrix for qualification testing has been established. Three of three full batches of irradiated cladding hulls have been cast. Spiked and cold samples castings are complete, and waste qualification testing has started. Glass-bonded sodalite is the ceramic waste form. Initial uranium and plutonium studies are available. Nonradioactive demonstration-scale equipment testing is complete. Equipment has been installed in the Hot Fuel Examination Facility. Laboratory-scale samples containing plutonium for accelerated alpha decay tests have been fabricated. At the time of the meeting (July 21, 1999), 4 of the 10 demonstration-scale cans had been processed.

A number of reports have been or will be produced by ANL relating to the demonstration project. Overall demonstration reports include *Spent Fuel Demonstration Final Report, Production Operations for the Electrometallurgical Treatment of Sodium-Bonded Spent Nuclear Fuel, Analysis of Spent Fuel Treatment Demonstration Operations*, and *Uranium Disposition Options*. Treatment operation reports can be divided into three groups: overall treatment reports, driver treatment reports, and blanket treatment reports. The overall report will be the *Development of Cathode Processor and Casting Furnace Operating Conditions*. Treatment operation reports include *Process Description for Driver Fuel Treatment Operations* and *Development of the Electrorefining Process for Driver Fuel*. Blanket treatment reports include *Process Description for Blanket Treatment Operations* and *Development of the Electrorefining Process for Blanket Fuel*. Waste operation and qualification reports are divided into three groups: overall waste reports, ceramic waste reports, and metal waste reports. The overall reports include *Waste Form Qualification Strategy, Waste Form Acceptance Product Specifications, Waste Compliance Plan*, and *Waste Form Degradation and Repository Performance Modeling*. The ceramic waste reports include *Ceramic Waste Form Process Qualification Plan* and the *Ceramic Waste Form Handbook*. The metal waste reports include *Metal Waste Form Process Qualification Plan* and the *Metal Waste Form Handbook*.

The Environmental Impact Statement covers sodium-bonded fuel treatment. The EBR-II spent fuel treatment demonstration was limited. It includes 100 EBR-II driver assemblies containing 410 kg of highly enriched uranium and 25 EBR-II blanket assemblies containing 1,200 kg of depleted uranium.

Robert D. Mariani, ANL-W, spoke on driver electrorefining results. One hundred driver assemblies have been treated in the Mark-IV electrorefiner within three years. Before the repeatability demonstration, 72 assemblies were used in tests. During the repeatability demonstration, 12 assemblies were used. Twelve more assemblies were electrorefined with repeatability conditions. The remaining four assemblies were used in additional tests.

The driver electrorefining test parameters, as detailed in ANL-NT-112, include anodic conditions; electrode configurations; the electrorefining sequence; cell current/cell voltage profiles; the uranium source; electrode, electrolyte, and cadmium agitation; and the Mark-IV electrorefiner temperature. Repeatability conditions for the Mark-IV electrorefiner, as described in ANL-NT-111,[1] include an anode voltage of <0.4 V when using fuel dissolution baskets, a 3,200 to 3,450 Amp-hour (A-h) per anode assembly, a dual anode/serial cathode electrode configuration, the electrorefining sequence, stepped current profiles, a 5 rpm electrode speed, a 20 rpm cadmium mixer speed, and a temperature of 500 °C.

The dual anode/serial cathode configuration combines two anode assemblies (four driver assemblies) and two cathode assemblies. The electrorefining sequence is direct transport number 1, direct transport number 2, followed by dissolution to the cadmium pool and deposition from the cadmium pool. There is a stepped current profile in direct transport number 1, with an A-h range from 0-480 A-h to 480-1,680 A-h to 2,700-3,500 A-h.

A summary of the repeatability demonstration for the Mark-IV electrorefiner shows that fixed electrorefining process conditions were used and that over the 61 working days of the demonstration (November 14, 1998, to January 22, 1999) the average treatment rate was approximately 24 kg per month, including an average electrorefining rate of 0.06 to 0.12 kg/h.

In summary, 100 driver assemblies were treated in 3 years, a continuous throughput of 16 kg of driver fuel per month over a 3-month period with fixed process conditions was exceeded (48 kg of driver fuel was treated in 61 working days), and material balances for uranium, plutonium, neptunium, sodium, and fission products were consistent with experimental error. Moreover, the uranium and plutonium balances complied with DOE Nuclear Material Control and Accountability requirements (DOE order 5633.3B).

Brian R. Westphal, ANL-W, presented information on cathode processing and casting results. Over the course of the demonstration, the cathode processor treated 40 driver batches, 6 blanket batches, and 8 cladding hull batches. The casting furnace during the demonstration treated 40 driver batches, 6 blanket batches, and 7 metal waste batches. Typical cathode processor/casting batch quantities were as follows: (1) for the driver, dendrites: 18 to 19 kg, salt (cadmium): 3 to 4 kg, DU: ~35 kg, uranium product: ~50 kg, and (2) for the blanket: blanket material: 40 to 41 kg, salt: 6 to 7 kg, uranium product: ~33 kg. Cathode processor/casting operating conditions for the driver and blanket include the following: for the cathode processor, the maximum crucible temperature is 1200 °C, the operating pressure is 0.1 torr, and the salt distillation step takes place over 1 h at 1100 °C. During casting, the maximum crucible temperature is 1300 °C, the operating pressure is at 900 torr until cast, and there is one stir cycle.

In summary, the driver demonstration was completed and the success criteria for these steps were met. One hundred driver assemblies were processed, 56 kg of uranium were processed during the repeatability step, condensate was returned during repeatability, operating conditions for the cathode processor/casting were specified, and three metal waste batches were cast. The blanket demonstration was initiated at the time of the meeting (July 21, 1999).

James L. Willit, ANL-W, spoke about electrorefiner throughput studies. The Mark-IV throughput goals for the demonstration phase were to electrorefine 150 kg of uranium in 1 month. Both glove box and hot cell tests

---

[1]R.D. Mariani, D. Vaden, B.R. Westphal, D.V. Laug, S.S. Cunningham, S.X. Li., T.A. Johnson, J.R. Krsul, and M.J. Lambregts, *Process Description for Driver Fuel Treatment Operations*, NT Technical Memorandum No. 111, Argonne National Laboratory, Argonne, IL, 1999.

demonstrate that this rate can be achieved. For treatment of the inventory of blanket fuel, the goal was to electrorefine 450 kg of uranium per month. The required average production rate has been demonstrated in glove box tests.

A process cycle was developed for sustained operation of the Mark-IV anode-cathode module (ACM). This process consists of loading, a stripping step, a washing step, a deposition step, another washing step, and then return to the stripping step, unless the product collector is full.

The Mark-V ACM stripping step removes the dense layer of uranium that builds up on the cathode tubes and stalls anode rotation. Reversing polarity electrotransports the dense layer of uranium from the cathode back to the surface of the anode baskets. The stripping step is terminated when all of the dense layer has been removed, as evidenced by a sharp increase in voltage. Periodically, low-current stripping is needed to fully clean the cathode tubes.

The washing step dislodges electrodeposited uranium trapped in the electrotransported zone, especially on the anode surfaces. Higher rotation speeds and multiple changes in direction of rotation improve the efficiency of this step. Implementation of this step decreases the frequency of stalls.

The deposition step's purpose is to electrotransport uranium from the fuel in the anode baskets to the cathode surface. The deposition step shows two voltage plateaus. There is a lower voltage plateau due to electrotransport of uranium from the basket surface back to the cathode. There is also a higher voltage plateau due to electrotransport of uranium from clad fuel segments inside the anode baskets. Electrodeposited uranium is scraped off the cathode surface and falls down into the product collector.

For throughput measurements in the Mark-V ACM, the key parameter is the average production rate, measured in grams of uranium per hour per ACM. The components of the average production rate include the net ampere hours per kilogram (338 A-h/kg being theoretical), which reflects the efficiency of electrotransport of uranium from the chopped fuel pins to the cathode tubes, and the shorting and U^{4+}/U^{3+} parasitic reaction decrease in coulombic efficiency and the increase in net A-h/kg. A second component is the stripping A-h/total A-h ratio. Minimizing this value has a huge impact on throughput. Additional components of the average production rate include the frequency and duration of low-current strips and the duration of the washing step.

A number of parameters affect the average production rate. For the deposition and stripping steps, these include the current and voltage, the anode rotation speed, and the cutoff voltage (resistance). For the washing step, parameters include the duration of the washing step, the rotation rate during the washing step, and changes in the direction of rotation. Other parameters include the cutoff voltage (resistance) and frequency of the low-current strips, and holdup and shorting.

Holdup and shorting always decrease throughput by decreasing coulombic efficiency (>338 A-h/kg uranium) and/or by increasing the stripping ampere hours. Product holdup has been found on the surfaces of the anode baskets, on isolated hard patches on the cathode tube, and in the region of the cathode supports. Cathode holdup must be removed by periodic low-current stripping. Anode holdup must be removed physically by hand or by aggressive washing.

Optimization of deposition parameters can be achieved in a number of ways. Higher deposition current favors higher throughput, where 600 A is the practical upper limit. Exceeding 200 A-h as the length of the deposition step results in a greater frequency of stalls. A higher voltage cutoff allows longer operation at higher currents. The present cutoff is 0.75 V, which avoids appreciable corrosion of the anode baskets. Lower rotation speed favors higher throughput but can increase the frequency of stalls. A range of 20 to 60 rpm has been examined, and additional optimization steps are in progress.

Optimization of stripping parameters can also be achieved in a number of ways. Lower rotation speed favors higher throughput but can increase the frequency of stalls. A range of 20 to 60 rpm has been examined, and additional optimization tests are in progress. A higher stripping current favors higher throughput. This decreases the length of the stripping cycle. A current of 600 A is the practical upper limit.

Optimization of washing step parameters was also accomplished in a number of ways. A shorter washing step increases the average production rate but can increase the frequency of stalls. The rate typically ranges from 1 to 6 min. A higher rotation rate decreases the frequency of stalls. The optimal rate was found to be 60 rpm. Multiple rotational direction changes were found to decrease the frequency of stalls. A forward-reverse-forward sequence was found to work better than a constant direction of rotation.

Other parameters were looked at as well. For product collector harvesting, removal by an inverted bake-out is slow but leaves little residue in the product collector. Removal by grinding is fast but leaves residues in the product collector. A higher concentration of uranium in the salt was found to increase the limiting current density. Lower concentrations favor a finer deposit morphology.

A number of comparisons were made between glove box and hot cell Mark-V ACM tests. Changes in operation parameters in general have the same qualitative effect in hot cell and glove box tests. Hot cell ACM tests show a somewhat lower average production rate (g/h/ACM). This is due to a higher stripping A-h/total A-h ratio in the hot cell tests. Holdup, shorting, and differences in adhesion of material to the cathode surface are the most likely sources for this difference.

A strategy was adopted for increased throughput. The deposition current was increased from 500 to 600 A. To decrease holdup and shorting, the stripping A-h/total A-h ratio was decreased, as was the A-h/kg rate. Finally, to improve handling operations, the uranium was ground out of the product collection reservoir rather than baked out.

Steven Sherman, ANL-W, presented information on blanket processing results in the hot cell. The setup of the Mark-V electrorefiner for blanket fuel is as follows. The fuel is depleted uranium at a rate of ~0.2 percent burnup. A 25-in. electrorefiner design of similar configuration has also been developed. The Mark-V contains four ports, each of which measures 10 in. and each of which operates independently. Each port has its own power supply, with a current of up to 600 A per port.

The experimental goals of the Mark-V electrorefiner were to maximize the production rate, to achieve simultaneous operation of the ports, to achieve a production rate of 150 kg of uranium per month, to realize reliable operation, and to perform unattended operation while passing current.

Significant achievements for the Mark-V electrorefiner were as follows. At the time of the meeting (July 21, 1999), the latest run conditions allowed 190 to 240 g of uranium/h/ACM as the average production rate, with 60 percent equipment utilization. There were four operational ports. Simultaneous operation of three ports was achieved, and routine operation of two ports was possible. More than five blanket assemblies were treated. Control software allowed unattended operation. Two product collector harvesting methods were developed: a bake-out oven, which gave a gravity-assisted product dump at 500 (C, and a product collector harvesting tool, a rotating multibladed tool for grinding out product at room temperature.

For Mark-V electrorefiner operation, the anode basket was loaded with 9 or 10 assemblies. Each assembly has 19 elements, and there are two anode baskets per assembly. Three product collectors are needed per anode basket. The operation is cyclic—each cycle is composed of deposition, stripping, and wash steps. The deposition step (per cycle) consists of a controlled current, initially 500 A for 200 A-h or until the 0.75-V cutoff is reached. If the voltage cutoff is reached before 200 A-h, the current is reduced and the step is continued. The rotation rate is 20 rpm, and the direction is forward. For the stripping step (per cycle), the current is 600 A, 300 A, and 165 A, each operating to a –0.75-V cutoff. The rotation rate is 10 rpm, and the direction is forward. For the wash step (there are two per cycle) the rotation rate is 60 rpm for 1 to 2 minutes in the forward direction. The end point of a batch is reached when the voltage is 0.75 at 80 A during deposition.

Materials/coatings tests were performed for intermediate containers. Alternative materials looked at included quartz, Pyrex, and aluminum oxide. A number of coatings on mild or stainless steel crucibles were also tested. Plasma-coated materials included zirconium oxide, zirconium oxide with a nickel-aluminum bond, tantalum, and tungsten. Titanium nitride was tested using chemical vapor deposition. Finally, boron nitride was tested by painting.

For each of these materials, the Mark-V product was placed into crucibles and heated to 500 °C for 2 to 3 hours. The crucibles were then removed from the furnace and cooled, and turned upside down to test for product adhesion. In all cases, the product failed to release at room temperature but released at 500 °C. Test results and metallography examinations suggest no chemical interaction between uranium metal and the crucibles. Tests with titanium nitride (conductive coating) were in progress at the time of the meeting (July 21, 1999).

At the time of the meeting (July 21, 1999), future work included the completion of the demonstration of 150 kg/month using established conditions. Run cycles were adjusted as needed to increase process reliability and production rates. Future equipment modifications to the ACM and support equipment were to be tried to increase process reliability and production rates. Testing of alternative materials and coatings and studies of the electro-refining characteristics of uranium were also to be continued.

Harold McFarlane, ANL, spoke about uranium disposition options. Disposition of uranium is constrained by the quantities and characteristics of uranium produced by treatment of sodium-bonded spent fuel. Options for disposition of uranium are constrained and defined by overarching DOE plans. These are discussed in the DOE publications *Disposition of Surplus Highly Enriched Uranium*, *Strategies for the Long-Term Management and Use of Depleted Uranium Hexafluoride*, and *Storage and Disposition of Weapons-usable Fissile Materials*.

Treatment of the remaining 3.2 MTHM of EBR-II driver fuel results in 10.7 MT of low-enriched uranium (19 percent ^{235}U). Fifty-five MT of depleted uranium (DU) would be recovered from treatment by EMT of the remaining blanket fuel in the DOE sodium-bonded SNF inventory. Approximately 1.6 MT of uranium ends up in the waste forms (0.89 MT in the metal waste form and 0.67 MT in the ceramic waste form).

Available options for uranium disposition include commercial use, continued storage, disposition as low-level waste (LLW), disposal as transuranic (TRU) waste, or other DOE use.

Excess highly enriched uranium (HEU), according to the *Record of Decision for the Disposition of Highly Enriched Uranium Final Environmental Impact Statement* (August 1996), will be dispositioned by one of two means: either conversion to low-enriched uranium light water reactor (LWR) fuel or disposal as LLW after blending down to 0.9 percent ^{235}U. The HEU in DOE spent nuclear fuel (SNF) is considered part of the excess HEU inventory.

To process HEU into LWR fuel for disposal as commercial reactor fuel, the blended product must be able to meet LWR fuel specifications. Standard fuel specifications are defined by ASTM. Specifications for off-spec fuel have also been developed for reactors operated by the Tennessee Valley Authority.

For disposal as LLW, the HEU needs to be blended to 0.9 percent ^{235}U. The key specification for disposal as LLW is that the TRU activity must be less than 100 nCi/g. After blending to 0.9 percent ^{235}U, the LEU product would contain approximately 40 nCi/gram of TRU activity.

Some fraction of the U.S. excess plutonium will be disposed of by can-in-canister. Current plans are to add plutonium-bearing LEU to the ceramic. Downblending will take place to 5 percent enrichment. Ceramic pucks will be fabricated in a remote glove box line. The feasibility of this option for the disposition of uranium is dependent on reducing the radiation levels of the metal ingots.

At present LEU cannot be directly converted to LWR fuel. Options are being explored to reduce key contaminants. Downblending processes are being considered that would improve the quality of the final product. Continued storage is feasible. More than 250 MTHM is being stored at ANL-W. Disposal as LLW is also feasible. Integration into the plutonium disposition program may be feasible.

Consideration is also being given to disposal of depleted uranium as low-level or transuranic waste. The TRU concentration in the DU ingots may be >100 nCi/g. ANL will determine whether it can lower this concentration to allow its classification as low-level waste. From a review of the waste acceptance criteria for the Waste Isolation Pilot Plant (WIPP), disposal as TRU waste appears to be feasible.

According to the *Final Programmatic Environmental Impact Statement for Alternative Strategies for the Long-Term Management and Use of Depleted Uranium Hexafluoride* (DOE/EIS-0269), the preferred alternative for management of DU is continued storage after conversion to oxide or metal. As part of this program, DOE will also support the development of markets for the oxide and metal DU products.

The DU produced from blanket treatment is a small fraction of that in the DOE complex (about 55 MT of 600,000 MT). Storage of 55 MT of recovered DU would be consistent with the programmatic Environmental Impact Statement. The material will already be a metal, which is the form most desirable for commercial application.

Depleted uranium disposition as an LLW is questionable without additional treatment. Disposition as TRU

APPENDIX B   91

metal waste may be feasible, with conversion to oxide as a fallback position. Storage and commercial uses are feasible.

Options for uranium disposition are constrained by DOE programmatic decisions based on environmental impact analyses. Within those constraints, there are viable options for both the LEU and DU from spent fuel treatment.

Sean McDeavitt, ANL, presented information on ceramic waste process and material studies. Salt-borne waste in the molten lithium chloride-potassium chloride electrolyte include transuranic isotopes, active fission products (e.g., cesium, strontium, barium, and iodine), and rare earth fission products (e.g., lanthium, cerium, and neodymium). Salt-borne wastes in the ceramic waste form include salt incorporated in the aluminosilicate (zeolite 4A) lattice. Zeolite transforms to sodalite during processing. The waste form is a glass-bonded sodalite composite.

Long-term thermal exposure of the ceramic waste form (CWF) was performed to assess the durability of the waste form. Five hot isostatic pressing (HIP) cans with reference ceramic waste composition were tested. Post-test examination was performed by leach testing, X-ray diffraction, microscopy, and bulk density determination. In summary, there were no major effects at 500 °C.

When looking at uranium and plutonium, the principal question is whether reactive salt components will react with the zeolite structure. There is concern about uranium chloride and plutonium chloride (UCl_3 and $PuCl_3$), and the rare earth chlorides. Understanding continues to improve. UCl_3 is known to react with water in the zeolite. $PuCl_3$ reaction with water is implied. Dried zeolite contains more than enough water to react with all the UCl_3 and $PuCl_3$ in the spent electrorefiner salt. Uranium does not appear to react with the lattice. Plutonium behavior is still being investigated.

Test methods for uranium include differential scanning calorimetry (DSC), which is the standard analysis tool. Post-test X-ray diffraction (XRD) is used to identify reaction products. DSC with simultaneous X-ray diffraction is a custom apparatus used at the Advanced Photon Source. X-ray diffraction is performed in situ during the DSC experiment. Evolved gas analysis (EGA) is also used. In this test the sample is heated in a vacuum chamber connected to a mass spectrometer. The evolved gas species are then identified.

In a high-uranium-content, heavy-salt mixture (i.e., lithium chloride-potassium chloride-UCl_3 with ~52 wt % uranium and 30 weight percent salt + 70 wt % zeolite), enough UCl_3 is present to consume the water in dried zeolite and still react with the zeolite lattice. DSC revealed the nature of the reactions. A small exothermic reaction begins at ~180 °C, followed by a strong exothermic reaction at ~350 °C. The reaction product contains uranium dioxide (UO_2).

Solid phases in the reaction were tracked by DSC/XRD. It was found that the zeolite lattice contraction begins at ~180 °C and that UO_2 formation begins at ~350 °C. Above 350 °C, the zeolite peak intensity decreases. Similar changes are also observed for lithium chloride-potassium chloride with and without UCl_3. Lattice reaction products are notably absent. EGA confirmed the reaction with water by detecting hydrogen chloride and hydrogen vapor evolution at ~350 °C.

These data indicate that uranium lattice reaction is not a problem. At high uranium concentrations, the zeolite lattice is not attacked by the salt. This implies that at low concentrations, the salt will not attack the zeolite lattice during waste processing. In addition, the salt cannot get past the water. These data also indicate that reaction of UCl_3 with water is a confirmed fact, and that the waste form will contain oxidized uranium particles. The impact of these particles on repository performance is being addressed through qualification testing.

There are a number of test methods to analyze plutonium in the CWF. X-ray diffraction has been used for blended powders contacted with plutonium-bearing salt. X-ray diffraction has also been used for hot, uniaxially pressed (HUP) ceramic waste form samples in powder form. X-ray absorption fine-structure (XAFS) spectroscopy has been used for HUP waste form samples in bulk and powder form. Electron microscopy has been used for ^{238}Pu-bearing HUP ceramic waste form samples.

A review of the data on plutonium in the CWF gave the following information. X-ray diffraction revealed that the zeolite lattice is unaffected during small-scale blending. It has also found that plutonium dioxide (PuO_2) is formed during small-scale blending. XAFS revealed information about the local plutonium environment. Plutonium was found to be present as PuO_2, not as $PuCl_3$. The PuO_2 particle size was estimated to be at least 20 to 40 Å. This indicates that the PuO$_2$ could not be inside the 12 to 16 Å zeolite cage. Electron microscopy revealed direct evidence of PuO$_2$. PuO$_2$ particles were observed near the sodalite-glass boundaries. Fine particles were observed by tunneling electron microscopy outside the sodalite.

Uranium and plutonium matrix samples were prepared to address questions about the CWF. The waste forms were evaluated using representative processing methods and compositions. Samples were prepared with reference fission product loading. The samples were subjected to heating blending and HIP. Two samples were prepared that contained plutonium- and uranium- bearing salts (1.5 mole percent PuCl$_3$ and 0.5 mole percent UCl$_3$, and 0.5 mole percent PuCl$_3$ and 1.5 mole percent UCl$_3$). Two samples were prepared with water in the zeolite (0.12 weight percent water and 3.5 weight percent water). Finally, four small HIP cans were prepared for each salt/zeolite combination. Analysis of the blended materials and salt precursors indicated low free chloride content after blending (0.03 to 0.3 percent). Also, low plutonium release was observed in the free chloride test solution (parts per thousand levels). X-ray diffraction of the blended material was under way at the time of the meeting (July 21, 1999). X-ray diffraction of the salt precursors was pending at the time of the meeting (July 21, 1999). Electron microscopy details were deferred to a later presentation.

A post-HIP test matrix was under way. The tests will evaluate the behavior of actinide-bearing phases to assess their impact on repository performance.

Pressureless consolidation has been looked at as a potential alternative to HIP. Preparation steps between pressureless consolidation and HIP are nearly identical. Complex HIP steps, including HIP can welding, powder compaction, evacuation, sealing, and welding, are eliminated. In addition, no high-pressure equipment is used in the hot cell with pressureless consolidation. The process is simple, fast, and compact. The blended powder is poured into settlers and the top surface is leveled. Loaded settlers are passed through a tunnel kiln at 850 °C for ~4 hours. The waste form is then removed and the settler is reused.

ANL is approaching the decision point for pressureless consolidation. The product is similar to the HIP product, without the can. The product has a uniform external appearance, and corrosion tests show no penalty in waste form leach behavior. The process also appears to be very easy to scale. Samples up to 1 kg have been prepared, and at the time of the meeting (July 21, 1999), lab scale-up to 5 kg was under way. The materials used are all available commercially. The graphite settler material is reusable. Finally, for the same cycle time, throughput may be more than 15 times higher than for HIP.

K.M. Goff, ANL-E, spoke on ceramic waste demonstration processing results. A number of nonirradiated CWF tests were summarized. The drying cycle for a 30 to 40 kg batch has been developed and implemented at an outside vendor. Thirty batches of material have been processed. Material from salt processing (crusher-mill/classifier) has been well characterized to obtain appropriately sized material for mixing. Twenty-two V-mixer experiments were performed, including one 80 kg test. The ability to produce salt-loaded zeolite with low free chloride has been demonstrated repeatedly. Ninety-four HIP cans have been produced, and the reliability of the HIP, the HIP can, the bake out/evacuation system, and the weld have all been demonstrated. It has been verified that the glass-zeolite ratio is maintained throughout all process steps. The durability of the waste form containing surrogates has also been well demonstrated.

Operating parameters for zeolite drying were as follows. The operating temperature was 550 °C, with a heating rate of ≤3 °C per minute. Time at temperature was seven hours, the degree of vacuum was ≤100 torr, the moisture content was 0.3 (≤1 weight percent), and rehydration was ≥18 weight percent.

For the salt crusher and mill/classifier, the mill speed was 1,500 rpm, the classifier speed was 350 rpm, the feed frequency was 7.0 Hz, and secondary gas readings were ΔP 0.8 in.

For the V-mixer, the operating temperature was 505 °C, time to temperature was ~2 hours (with no limits for heater ramp rates), and the time at temperature was 15 hours. The rotation rate was 17 rpm, free chloride measured ≤ 0.5 weight percent, and salt content was 3.8 Cl ions per unit cell.

For demonstration runs using HIP, loading was as follows. Twenty-one bellows cans were loaded with 1,600 g of material. ANSTO cans were loaded with 950 g of material. The witness tubes were loaded with 24 g of material. The HIP cans themselves were made of 304L stainless steel. The compactor operating time was between 0 and 10 minutes.

APPENDIX B 93

Parameters for HIP evacuation bake-out and crimp and weld were as follows. For evacuation/bake-out the vacuum was at <100 mtorr, furnace time at temperature was ≥6 h, and the furnace temperature was 500 °C. For crimp and weld, the magnitude of the crimp was 12 tons, and the type of weld was tungsten inert gas (TIG).

Eleven nonirradiated HIP cans were processed through the HIP in the Hot Fuel Examination Facility (HFEF). At the time of the meeting (July 21, 1999), processing of irradiated materials was ongoing.

Operating parameters for the hot isostatic press are as follows: the hold temperature is 850 °C, the maximum pressure is 14,500 psi, the temperature ramp is 5 °C per minute, the pressurization rate is 240 psi/min on average, the depressurization rate is 85 psi/min on average, the hold time is 60 min, the shelf time is 45 min, and the shelf temperature is 750 °C.

Demonstration-scale irradiated samples were produced. Irradiated salt was removed from the electrorefiner and transferred to the HFEF. Mill/classifier tests were performed to confirm the behavior of the electrorefiner salt. A batch of 100-driver salt was successfully processed through the V-mixer. At the time of the meeting (July 21, 1999), irradiated samples and witness tubes were being processed through the HIP.

In the HIP scale-up tests, no density variations were detected from 27 samples taken throughout the can. X-ray diffraction was performed on 13 samples, and the diffraction patterns were the same as seen in the demonstration products. Initial performance results were the same as those obtained at the demonstration scale. An 18-in. can was processed in June 1999. This diameter is essentially production scale. Hold time was approximately 10 h. The can held 60 kg of material. An 18-in. ANSTO can was also processed. This can held almost 100 kg of material.

Samples were produced for accelerated alpha decay studies. Two batches of material containing ^{238}Pu were produced, and characterization is ongoing. The salt contained 15 mole percent plutonium and surrogates for fission products. The HUP sample contains more than 2 weight percent plutonium.

Results of the alpha decay studies show no significant changes since last reporting. The plutonium normalized release rate was 1×10^{-4} g/(m² day). There were no changes in density. There were also no changes in the unit cell volume of sodalite.

L.R. Morss, ANL-E, spoke about CWF qualification testing. The CWF is ~75 percent sodalite, ~25 percent glass, with small amounts of UO_2, PuO_2, oxychlorides, halite, and nepheline. Qualification needs that are addressed in the testing program cover many areas. Repository performance assessment needs require the characterization of matrix corrosion behavior and radionuclide release, and also require a mechanistic model with model parameters. Waste specification needs require the identification of phases containing radionuclides, and methodology must be provided to monitor product consistency. For process qualification, a methodology must be provided and a database established for process operating parameters.

Scoping tests identify corrosion modes and provide model parameter values. Solution exchange tests demonstrated that the primary release mechanism is matrix dissolution, not leaching of occluded material. Product consistency tests characterize the dissolution behavior under concentrated solution conditions. The MCC-1 static leach tests characterize dissolution behavior under dilute solution conditions and provide model parameter values. The pH buffer tests measure pH dependence of the dissolution rate. Accessible free salt measurements measure the amount of soluble salt (mostly sodium chloride).

The 7-day product consistency test is specified in the waste acceptance systems requirements document (WASRD) for defense high-level waste (DHLW) glasses.[2] Crushed material in demineralized water at 90 °C is used. The high surface area of material/volume of solution (S/V) ratio is representative of anticipated Yucca Mountain conditions. ANL has shown that the performances of the reference CWF and DHLW glasses are similar in this test. Longer-term tests are used to study corrosion behavior.

The normalized elemental mass loss is predicted by the following equation:

$$NL(i) = C(i) \, (S/V) \, f(i)$$

---

[2]American Society for Testing and Materials, *Standard Test Methods for Determining Chemical Durability of Nuclear Waste Glasses: The Product Consistency Test (PCT)* Standard C1285-94, Annual Book of ASTM Standards, Vol. 12.01, West Conshohocken, PA.

where C(i) is the measured solution concentration, S/V is the surface area of material divided by the volume of solution, and f(i) is the mass fraction of the element in the material. This equation allows direct comparison of the release of different elements from different materials and at different S/V ratios.

The MCC-1 test was developed in the early 1980s to evaluate the relative chemical durabilities of simulated and radioactive monolithic waste forms at low S/V ratios. Monolithic samples in demineralized water at 90 °C are used. The low S/V ratio is not representative of anticipated Yucca Mountain conditions. The low S/V ratio allows the MCC-1 test to be useful for determining the initial (forward) rate and is thus useful for providing modeling parameters. MCC-1 tests found that the response of reference CWF was similar to that of DHLW glasses.

In the scoping tests, the dissolution behaviors of sodalite, the glass binder, free salts, and oxides/oxychlorides were determined. The dissolution of free salts becomes controlled by the dissolution of sodalite and glass binder. The dissolution behaviors of sodalite and glass binder phases can be described with the glass dissolution model in the total systems performance-viability assessment. Model parameter values for DHLW glass provide the upper bound to parameter values for the sodalite and glass binder phases.

For the corrosion model, a rate expression was developed for aluminosilicate minerals:[3]

$$\text{Rate} = k_f (1 - [H_4SiO_4]/[H_4SiO_4]_{sat}).$$

The rate depends on orthosilicic acid concentration, but not on time. The value for k_f depends on temperature and pH. The value for $[H_4SiO_4]_{sat}$ depends on temperature. The equation has been modified for application to DHLW glasses:[4]

$$\text{Rate} = k_f (1 - [H_4SiO_4]/[H_4SiO_4]_{sat}) + k_{long}.$$

The rate expression is:

$$R = k_0 10^{(\eta \bullet pH)} e^{(-E_a/RT)} (1 - [H_4SiO_4]/[H_4SiO_4]_{sat}) + k_{long}$$

where R is the alteration rate of the waste form (g/(m²d)), k_0 is the intrinsic rate (g/(m²d)), η is pH dependence, E_a is temperature dependence (kJ/mole), $[H_4SiO_4]$ is the concentration of H_4SiO_4 in solution (mg/l), $[H_4SiO_4]_{sat}$ is silica saturation concentration (mg/l), and k_{long} is the residual long-term rate (empirical term). Tests showed that $[H_4SiO_4]_{sat}$ is lower for sodalite, the glass binder, and the CWF than for HLW glasses.

Implementation of the CWF model in the repository integration program uses radionuclide inventories and the specific surface area of the CWF. The HLW glass rate parameters are currently used by ANL to model CWF behavior. Experimentally determined rate data for the CWF will be included in future models.

Scoping tests have identified the primary corrosion mode as matrix dissolution. They have also characterized short-term and long-term corrosion. The scoping tests have also shown that the CWF corrosion follows the model for aluminosilicate minerals and HLW glasses. For the model parameters, pH dependence of the forward dissolution rate is now being determined for sodalite and glass. Temperature dependence will be determined. The saturation concentration of H_4SiO_4 is being determined. A simplified model has been implemented in the Repository Integration Program (RIP) program.

At the time of the meeting (July 21, 1999), work in progress included tests to study release behavior of PuO_2 from plutonium-loaded CWF. Tests to determine the temperature coefficient of the forward rate were under way. Tests to determine if the aluminum concentration in solution affects the dissolution rate had begun. Tests were also initiated under repository-relevant conditions (drip tests, vapor hydration tests).

---

[3] P. Aagaard and H.C. Helgeson, *Am. J. Sci.* 282, 1982, pp. 237-285.
[4] B. Grambow, *Mat. Res. Soc. Symp. Proc.* 44, 1985, pp. 15-24.

S.G. Johnson, ANL-W, spoke about ceramic waste uranium/plutonium behavior studies. The approach is to prepare the CWF with uranium, plutonium, and fission products by blending and HIP. Two uranium/plutonium ratios are used, as are two water contents in the zeolite. The microstructures of these materials are then analyzed, corrosion tests are carried out, and the disposition of uranium and plutonium released from the CWF is determined.

The main objective of this work is to determine uranium/plutonium release as the CWF matrix corrodes. This includes determining uranium and plutonium released into solution, uranium- and plutonium-containing colloids in solution, if any, and alteration phases containing uranium and/or plutonium.

The purpose of the corrosion test matrix is to achieve accelerated corrosion conditions. Here also the goal is to determine release of uranium and plutonium into solution. Also, altered solids remaining with the CWF are characterized. Plutonium colloids are also characterized in solution by sequential filtration, dynamic light scattering, and transmission electron microscopy (TEM). A product consistency test (PCT) at 120 °C in demineralized water at S/V ratios of 1,000 to 2,000 m^{-1} was also performed.

Corrosion tests were also carried out at reference conditions. This was to verify that accelerated conditions do not change the release mechanism. PCT at 90 °C was performed in demineralized water at S/V 2,000 m^{-1}. The approach to repository relevant conditions was to perform drip tests with cores from the CWF.

Physical characterization of the CWF entailed determination of the coordination and distribution of uranium and plutonium. This included structural identification of uranium/plutonium-containing phases, location of uranium/plutonium-containing phases in the waste form, and distribution of phases within HIP cans (radially and axially). Phases were also identified in the precursors (zeolite and salt-loaded zeolite), glass-bonded sodalite (CWF), solid reaction products after corrosion tests, and colloids present in solution after corrosion tests.

Physical characterization tests looked at plutonium before, during, and after the corrosion tests. Before the corrosion tests, plutonium was looked at in the CWF phases. This was done by X-ray powder diffractions, X-ray absorption fine structure spectroscopy (XAFS), scanning electron microscopy (SEM), and transmission electron microscopy (TEM). During the corrosion tests, plutonium was released into solution. This was observed by sequential filtration to estimate the size and composition of colloidal particles. Dynamic light scattering determined the size distribution of colloidal particles. TEM looked at colloidal particles wicked onto the grid. Plutonium was also looked at in the alteration products of corrosion tests. SEM and TEM were used on the surfaces of reacted solids after the corrosion tests.

Initial observations from the analyses are as follows. Plutonium and uranium are found in inclusions with rare earths in the form of solid solutions of oxides, silicates, or a mixture of these two. The plutonium/uranium oxide phase is small in size, typically ~10 nm, while the silicate phase is larger, typically 50 to 1,000 nm. in size. Most actinide inclusions are in the glassy regions near sodalite grains. More actinide inclusions are present in the sodalite granules in the waste forms produced with the dry zeolite than in those produced with the wet zeolite. The uranium/plutonium ratio does not affect the overall microstructure of the waste form.

SEM analysis observations include the following. Plutonium, uranium, and rare earths are present as oxide solid solutions and silicate solid solutions. Actinide-bearing crystals accumulate mostly in the glassy regions. Small amounts are observed in the sodalite regions (more in the dry samples than in the wet samples). The uranium/plutonium ratio does not affect the microstructure of the ceramic waste form.

TEM analysis showed that uranium, plutonium, and rare earths are present in silicate phases (50 to 1,000 nm size). In addition, (uranium, plutonium, rare earth) oxides are present in solid solution (~10 nm in size).

XRD observations show that the XRD pattern is similar to the reference CWF pattern. The plutonium/uranium oxide phase behaves as a solid solution. No significant variation was observed in the sodalite crystalline pattern between samples with low and high moisture content. The minor halite phase reflection is roughly twice as intense in high moisture samples as in low moisture samples.

Tests and analyses to study the release of uranium and plutonium from the CWF as colloids and dissolved species had been initiated and were in progress at the time of the meeting (July 21, 1999). The morphology and chemical composition of the uranium/plutonium containing phases were also being determined at the time of the meeting.

Thomas P. O'Holleran, ANL-W, spoke on ceramic waste product consistency testing. The goal of product consistency testing of the CWF is to ensure that the waste form production process is well controlled. A well-controlled process will create waste form products that meet ANL's waste acceptance product specifications and can be subsequently disposed in a geologic repository. A suite of tests are to be applied to make sure that the waste form is consistent from run to run.

A number of tests and methods were applied to the ceramic waste form for product consistency. These included density, X-ray diffraction, microscopy, immersion tests, and toxic characteristic leaching procedure (TCLP).

Density was examined because the degree of consolidation in the HIP cans is approximately twice as great in the pre-HIP stage as in the post-HIP stage. This consolidation was important and served to re-check that the HIP went through the appropriate temperature-pressure (T-P) cycle.

X-ray diffraction gave XRD patterns that revealed whether the salt-loaded zeolite 4A had completely converted to sodalite during the processing in the HIP. Conversion to sodalite was important to waste form durability and double-checked that the HIP had gone through the appropriate T-P cycle. X-ray also served as a check for the formation of any anomalous phases during production that could alter the release characteristics or other important properties.

Microstructure was an important indicator of the interaction of the feed materials with the environment and each other during the processing and the appropriate T-P cycle in the HIP. Scanning electron microscopy (SEM) gave information on grain size, relative porosity, and semiquantitative elemental composition of various phases. Transmission electron microscopy (TEM) gave information similar to SEM but on a much smaller scale. In addition, it gave electron diffraction for the determination of the crystalline structure of phases.

Immersion tests were used to assess the consistency of the product via the interaction with water at 90 °C. The release of pertinent elements from all significant phases present in the waste form were monitored. These included matrix elements silicon, aluminum, and boron; alkali/alkaline earth elements lithium, sodium, potassium, rubidium, cesium, strontium, and barium; chloride and iodide anions; and the actinides uranium, neptunium, and plutonium.

The toxicity characteristic leaching procedure (TCLP) was used because Yucca Mountain will not be licensed as a RCRA facility. The TCLP test to assess the potential escape of hazardous material from the sample. The hazardous character of the waste must be determined.

The tests/methods were put in place to monitor the CWF product to ensure its consistency, demonstrating a well-controlled process. These tests/methods are sensitive to density changes, phase composition, microstructure, release rate of important elements and possible release of hazardous constituents.

Daniel P. Abraham, ANL-E, spoke about metal waste qualification testing. The metal waste form test plan is based on an ASTM C 1174. At the time of the meeting (July 21, 1999), 99 percent of phase 1 testing (for the demonstration) was complete; the results are in the Metal Waste Form Handbook.[5] Phase 2 testing, which extends beyond the demonstration, had just begun at the time of the meeting (July 21, 1999). The focus was on experiments. This was to generate input for the total systems performance assessment (TSPA) analysis. These experiments are to specify the composition range for process qualification. The TSPA analysis requires input on corrosion behavior. This includes data feeds into corrosion models. These models will be incorporated into RIP performance assessment software. The analysis also requires input on physical properties and phase stability. The process qualification input requires information on composition-property relationships. A range of compositions was studied: 0-20 weight percent zirconium, 0-5 weight percent noble metals (with up to 2 weight percent technetium), and 0-11 weight percent actinides (uranium, plutonium, and neptunium). Alloy properties considered were microstructure, mechanical and thermophysical properties, and corrosion behavior.

---

[5]D.P. Abraham, S.M. McDeavitt, D.D. Keiser, Jr., S.G. Johnson, M.L. Adamic, S.A. Barker, T.D. DiSanto, S.M. Frank, J.R. Krsul, M. Noy, J.W. Richardson, Jr., and B.R. Westphal, *Metal Waste Form Handbook*, NT Technical Memorandum No. 121, Argonne National Laboratory, Argonne, IL, 1999.

Microstructural studies were performed on the MWF to identify phases and to study noble metal and actinide distribution. There were 102 tests performed, including SEM, TEM, and neutron diffraction. All tests had been completed at the time of the meeting (July 21, 1999). No noble metal-rich precipitates were observed. Actinides are present only in the intermetallic phases. It was also found that the as-cast microstructure is stable up to ~1100 °C.

A number of tests were performed on the physical properties of the MWF. For mechanical properties, tensile, compression, and impact tests were performed on the MWF. There were 45 tests using specimens from eight ingots. For thermophysical properties, density, thermal conductivity, specific heat, and the thermal expansion coefficient of the MWF were all determined. All tests had been completed.

Immersion tests provide information on the selective leaching of elements. Static immersion tests (90 °C and 200 °C, up to 1 year) were performed on 176 samples. All are complete. Forty-two pulsed-flow immersion tests (90 °C) were performed. At the time of the meeting (July 21, 1999), all were in progress. Results show good fission product and actinide retention for the MWF alloys.

Immersion tests at 90 °C show minor surface tarnish and negligible weight change. Fission product release at one year for technetium was 0.05 g/m^2. The maximum rate was ~ 6×10^{-4} g/m^2 per day. Losses of palladium, ruthenium, rhodium, and niobium were 0.01 g/m^2. Other elemental releases after one year include 0.2 g/m^2 for iron, chromium, nickel, manganese, and molybdenum. For uranium, the release was 0.2 g/m^2. The maximum rate for uranium was ~ 5×10^{-3} g/m^2 per day.

Electrochemical tests provided a relative measure of the corrosion rates for various compositions tested. Three hundred and sixty tests were performed, and all were completed. Vapor hydration tests provide information on the corrosion behavior in superheated steam (200 °C). There were 44 tests. At the time of the meeting (July 21, 1999), 36 were complete.

Electrochemical corrosion testing parameters included using samples containing a range of zirconium, fission product, and actinide contents. The pH of the solution was varied (pH = 2, 4, 9, and 10). MWF corrosion rates were similar for all alloys (even those with uranium and technetium). Rates were comparable for those for SS316 and Alloy C22 and were two orders of magnitude smaller than for mild steel. Similar results were seen for tests in pH = 4, 9, and 10 solutions. Rates for pH = 2 were higher.

Vapor hydration corrosion tests were performed in steam at 200 °C and 100 percent humidity. Test periods were for up to six months. The oxide layer thickness for the MWF samples were ≤1 μm for both 56- and 182- day tests. The adherent oxide layer appears to retard MWF corrosion. TEM and AES studies were performed to identify oxide(s) and to determine the corrosion mechanism.

For galvanic corrosion, electrochemical tests provide information on the galvanic interaction between waste forms and candidate container materials. The galvanic current is measured between the metallic samples and Alloy C22. Samples are short-circuited through a computer-controlled potentiostat, which acts as a zero-resistance ammeter. There were 10 tests, and all were complete. The steady-state current value for mild steel in J-13 (pH = 9) was 4 μA. At pH = 2, the value was 40 μA. For SS-15Zr in J-13 and pH = 2, the value was <1 μA. Galvanic interaction between the MWF and C22 was found to be insignificant.

At the time of the meeting (July 21, 1999), the MWF test plan was 99 percent complete. Alloy microstructures are well understood. The alloys show favorable mechanical and thermophysical properties and good corrosion resistance.

Dennis D. Keiser, Jr., ANL-E, spoke about metal waste product testing. Three primary issues surround the MWF. Have appropriate casting conditions been determined? What characterization techniques are employed for the MWF product? What are typical results from these techniques?

MWF casting results in the fuel conditioning facility (FCF) are as follows. Three MWF ingots in the FCF that can accommodate an amount of cladding hulls corresponding to that of two driver fuel assemblies have been cast. A fourth MWF ingot with blanket cladding hulls substituted for driver hulls has been cast. The bulk of EBR-II material left to be processed will consist of blanket elements. The conclusion is that appropriate casting conditions have been determined and implemented for both driver and blanket hulls.

For characterization of irradiated MWF ingots, the sampling strategy involved core drilling, where samples are representative of the final product, and many were obtained to establish the homogeneity of the final product.

Injection casting was used to obtain a sample to correlate bulk chemistry results with those obtained from core drilling. Analysis methods included microscopy, bulk chemistry, immersion tests, and toxicity characteristic leaching procedure (TCLP).

Observations from chemical analyses show that elemental analyses of core drilled samples compares well with those for injection-cast samples. The variability of the chemical analysis associated with the use of core-drilled samples has been established. The noble metal fission products technetium, ruthenium, and palladium are present in the MWF at tenths of a weight percent.

Observations from microstructural analyses show the expected amounts of alloy constituents present in the MWF alloys (i.e., stainless steel components, zirconium, uranium, and noble metal fission products). The constituents are partitioned between typical alloy phases. The major phases are ferrite and $Zr(Fe, Cr, Ni)_{2+x}$. The minor phases are austenite, Zr_6Fe_{23}, and uranium phases containing technetium and selenium. Noble metals and uranium favor specific phases. Technetium favors iron solid solution phases; ruthenium, palladium, and uranium favor intermetallic phases. The conclusion is that the results show good agreement with those for surrogate MWF alloys.

The purpose of immersion testing is to show the correlation between actual MWF samples and doped MWF samples. CFMW06 and 07 monolithic samples were tested in the following manner: core-drilled samples 0.8 cm in diameter × 1.3 to 2.3 cm thick were employed. The samples were cleaned ultrasonically several times in 200-proof absolute ethanol. Deionized water and J-13 well water were used as test solutions. The surface area/volume ratio equaled 50 m^{-1}. Stainless steel containment vessels were used as blanks. The vessels were heated to 90 °C for 2 weeks. Leachates were analyzed and acid stripping of the vessels was performed and analyzed. Immersion testing results for hot ingots were comparable to results obtained for cold samples.

The purpose of TLCP testing was to satisfy the waste acceptance system requirements document, which requires that a waste producer determine if the waste for disposal has hazardous characteristics. The method used was to core-drill samples from MWF ingots CFMW06, 07, and 08 and subject them to TCLP. TCLP tests for release of cadmium, chromium, lead, arsenic, selenium, silver, barium, and mercury utilized contact with mild or buffered acid. The solution was agitated at 18 rpm for 18 h. A solid:liquid ratio of 100 g to 2 liters was used. Solid waste that was tested had to pass through a 9.5-mm mesh. The conclusion of these tests was that the core samples from the three MWF ingots pass the TCLP release limits for the eight toxic metals.

The answer to whether appropriate casting conditions had been determined was affirmative. The three MWF ingots were cast in the FCF that incorporate an amount of cladding hulls corresponding to two driver assemblies, with the third MWF ingot having been cast using blanket hulls. The characterization techniques employed for the metal waste form product were bulk chemistry, SEM, an immersion test, and the TCLP analysis. The immersion test shows a durable waste form that compares well to cold surrogate alloys. The TCLP analysis indicates that the MWF is not a characteristic waste. The bulk chemistry confirms the anticipated composition. The microscopy identifies the structure and the disposition of the radionuclides to be immobilized.

Mark C. Petri, ANL-E, gave information about metal waste release modeling. The purpose of this modeling is to develop a radioisotope-release-rate model for the stainless-steel-based MWF. This in turn is used as an input module for the RIP, the performance assessment code. Another purpose was to assess experimental needs to support modeling beyond June 1999.

The MWF modeling approach uses known stainless steel (SS) degradation modes. These include uniform aqueous corrosion and crevice corrosion. A comparison is made between empirical correlations for SS corrosion and MWF data. Assumptions are made about the extent of crevice corrosion across the MWF surface. It is assumed that the corrosion rate gives the radioisotope release rate.

Results of the modeling effort for SS uniform aqueous corrosion rate data show that limited data are available for repository conditions. There is significant data scatter. The MWF and SS have similar corrosion rates. The data also show a correlation between electrochemical results and immersion results. The SS crevice corrosion rate data show that data are virtually nonexistent for repository conditions. One electrochemical study indicates a [Cl⁻] dependence.

An empirical corrosion rate correlation has been developed based on SS electrochemical data. The MWF is assumed to follow the same correlation. The MWF degradation model has been incorporated into the RIP performance assessment code. Additional MWF corrosion experiments are planned to support modeling.

Regression of 316 SS electrochemical uniform aqueous corrosion data shows two key variables: pH and the product pH•T. For 316 SS, the chloride content has no significant effect on uniform corrosion. For 304L SS, passivation breakdown is observed beyond 100 mg/l Cl^-.

Results for MWF uniform aqueous corrosion data were reported. In simulated J-13 water at room temperature, pH = 4 gave a value of 88 mg/l Cl^-, while at pH = 2, the value was 443 mg/l Cl^-. MWF corrosion rates were independent of composition within the specified composition range (zirconium, technetium, uranium, etc.). The MWF has corrosion rates similar to 316 SS. No corrosion rate jump was observed with Cl^- levels up to 443 mg/l, unlike 304L SS.

For 304L SS crevice corrosion, crevice corrosion rates are difficult to measure; very few data are available. There are no data on pH or temperature dependence. A previous 304L SS electrochemical study showed a logarithmic relationship between crevice corrosion rate and chloride and fluoride content (up to 800 mg/l). These are relative (not absolute) corrosion rates.

MWF corrosion rate modeling assumes that SS relationships hold for the MWF. Another assumption is a conservative 1:4 ratio between the active crevice area (anode) and the cathodic area. This ratio is the largest seen for SS. Immediate crevice corrosion initiation is also assumed. It is assumed that the crevice provides no cathodic protection, i.e., all regions undergo corrosion at either a uniform rate or at the crevice rate.

The net MWF corrosion rate model follows the equation

$$CR = 0.8\ CR_{uniform} + 0.2\ CR_{crevice}$$

$$CR = (10^{-0.182\ (pH) + 1.43E\text{-}3\ (pH)(T)})(94.2 + 0.0569\ [Cl^-]^{1.22})$$

where CR = the net corrosion rate over the MWF surface in $g/m^2/yr$, pH = the pH of the water in contact with the MWF (2 < pH < 10), T = the temperature of the MWF surface in °C (20 °C < T < 95 °C), and $[Cl^-]$ = the chloride (and fluoride) ion concentration of the water in mg/l ($[Cl^-]$ < 443 mg/l).

When looking at the net MWF corrosion rate model vs. chloride concentration, a noticeable increase in the predicted MWF corrosion rate beyond 200 mg/l Cl^- was observed. This was attributed to crevice corrosion.

E.E. Morris, ANL-E, spoke about repository performance modeling. The purpose of the Argonne performance assessment work is to evaluate the performance of ANL waste forms under repository conditions. This provided feedback on the effect of waste form dissolution parameters on performance in the repository. The performance assessment used a simplified version of the TSPA-VA RIP model. The simplified model permits calculations to be performed 10 times faster than with the TSPA-VA RIP model. It also facilitates parameteric studies. The model was constructed for ANL by Golder Associates Inc., the developer of RIP. Calculations show that the ANL waste form performance is comparable to the TSPA-VA waste form performance.

The simplified model used the same engineered barrier system model as the TSPA-VA model. Unsaturated and saturated zone radionuclide transport is represented by simpler models. This produces essentially the same dose rates at the 20-km well as does the TSPA-VA model. The expected-value time histories for dose rate hold, as do complimentary cumulative distribution functions (CCDF) for the peak dose rate.

Results for the 39- and 9-isotope model results were compared. The 39 radionuclides used in TSPA-95 are adequate to represent the ANL waste forms. The 9-isotope simplified model is needed because the 39-isotope model is too slow. The 39- to 9-isotope comparisons with no ANL waste are used to verify the understanding of how the actinide inventory adjustments were made for the TSPA-VA. The 39- to 9-isotope comparisons with only ANL waste are used to verify that actinide inventory adjustments are satisfactory for the ANL waste forms.

The 9-isotope total dose rate time history agrees well with the 39-isotope result. Several radionuclides that contribute a few percent to the total dose rate in the 39-isotope model are not included in the 9-isotope model. The general agreement within 15 or 20 percent between the total dose rate in the two models was considered satisfactory.

Conclusions of the parametric study were that the dominance of commercial SNF and the importance of its cladding credit, are evident for the TSPA-VA base case repository loading. In the absence of cladding credit, a waste matrix with a dissolution rate comparable to or slower than that for DHLW glass can significantly reduce the peak dose.

The RIP implementation is similar to that for DHLW glass. The model is currently the same as DHLW. The rate data are those from DHLW. The surface area per gram is $1.8 \times$ DHLW and is constant throughout the RIP calculation. The radionuclide inventory is that of ANL ceramic waste. ANL experimental data will be included in future models.

The impact of waste form dissolution rates was assessed by comparing time histories of the normalized cumulative release from the engineered barrier system. The impact on the repository was assessed by comparing complimentary cumulative distribution functions for the peak dose rate at the 20-km well.

In summary, a RIP modeling capability for ANL waste forms has been developed and demonstrated. The repository performance assessment effort is on schedule to meet the data call for Yucca Mountain project site recommendation/license.

Stephen G. Johnson, ANL-W, presented information on the waste qualification strategy at ANL. The goals were to comply with all pertinent legislation on production and ultimate disposal of the waste forms. Another goal was to provide comprehensive plans and adequate records to prove that the process is well behaved operationally and creates a consistent product that can be dispositioned in a geologic repository. A further goal is to consult with the cognizant branches of DOE about the spent fuel treatment process under development at ANL and the waste forms that will be produced.

Pertinent legislation on this topic includes Nuclear Regulatory Commission legislation 10CFR60. The NRC regulations in Part 60 of title 10 of the code of federal regulations provide that containment of radionuclides shall be substantially complete for a period that shall be no less than 300 years nor more than 1,000 years, unless otherwise permitted by the NRC. Any release of radionuclides after the containment period shall be a gradual release and limited to certain small fractional amounts based on the calculated inventory present at 1,000 years after closure. This stipulates that data on the waste forms must be gathered and integrated into the performance assessment of the geologic repository. Another pertinent law is Environmental Protection Agency (EPA) legislation 40CFR191,[6] which provides that cumulative releases of radionuclides from the disposal system for 10,000 years after disposal shall have a likelihood of less than 1 chance in 10 of exceeding the values stated for each radionuclide in the regulation. This also stipulates that data regarding waste form performance must be gathered and integrated into the performance assessment of the geologic repository.

There are a variety of data needs of the geologic repository for high-level wastes (HLW). An Environmental Impact Analysis is required to determine the quantity and type of material acceptable and to determine repository design. Performance assessments, including a viability assessment, site suitability, site recommendation, and license application are also needed. Two documents are related to waste acceptance criteria needs, the WASRD, and the mined geologic disposal system waste acceptance criteria (MGDSWAC).

The geologic repository is responsible for the total system performance assessment (TSPA). The objective of the waste disposal system is to contain and isolate the radioactive constituents so that the dose impact to humans is attenuated to a relatively benign level. Water contact is limited, the waste package is long lived, the release rate of radioactive constituents from breached waste packages is low, and a concentration gradient is set during transport of the radioactive constituents that are released from the waste packages.

As for how to provide the data required for the repository performance assessment, the National Technology Transfer and Advancement Act (Section 12, 1996) stipulates, "All federal agencies now must use technical standards developed or adopted by voluntary consensus standards bodies (SDOs) when carrying out procurements and policy objectives.... Federal agencies and departments are now expected to work with SDOs and to have their employees participate in the SDO's development of technical standards."

---

[6]Yucca Mountain is exempted from 40CFR191.

To provide the data required by the repository performance assessment, ASTM C1174 is used to provide guidance in the selection of the types of tests and methods to apply. These include attribute tests, characterization tests, accelerated tests, service condition tests, analog tests, and confirmation tests. ASTM C1174 provided the basis for developing the test matrices for the CWF and the MWF. These test matrices are extensive and will provide the data necessary to develop a model for assessing the performance of the waste forms that result from the spent fuel treatment process.

The WASRD stipulates chemical composition, radionuclide inventory, phase stability and integrity, and product consistency as requirements. The waste acceptance qualification specification (WAPS), the waste compliance plan (WCP), and the waste qualification report (WQR) are all required documents. The WAPS sets out specifications for the process, e.g., canister dimensions, radionuclide content, and hazardous characteristics. The WCP states specifically, in detail, how each specification in the WAPS will be demonstrated and how that compliance will be documented. WQR compiles the results from the waste form/process testing and analysis to demonstrate the ability of the producer to be in compliance.

Test matrices are organized to fulfill the broad data needs for waste qualification. Data needs include the performance assessment, the waste product specification, process qualification, and background tests.

Preliminary discussions between ANL and the DOE Office of Environmental Management, Office of Civilian Radioactive Waste Management, and Office of Nuclear Energy have taken place. Topics discussed included product specifications, compliance plans, and classification of electrometallurgical technology waste forms.

Waste qualification activities that were to be completed by the end of the demonstration include the test matrices, except for long-term tests and radioactive sample tests. Methods will be established for applying both waste form models to the repository time frame. Sensitivity performance studies will be performed using bounding degradation models. For both waste forms, degradation modeling refinements based on experimental data will be incorporated into the simplified TSPA model.

Other activities to be completed include the preliminary waste acceptance product specifications, which will be submitted to DOE for comment. Process parameters will be identified for full-scale operations qualifications Future waste test matrices for license application were to be established by June 1999.

Post-demonstration-phase waste qualification activities include the completion of waste form test matrices that support repository license application and are included in the WQR. The WQR was to be written. Waste process qualification tests with full-scale equipment will be performed.

JULY 22, 1999—CLOSED SESSION

The committee met in closed session on July 22, 1999 at the Shilo Inn in Idaho Falls, ID. Discussions consisted of the previous day's presentations by ANL personnel and preliminary discussion of the structure and writing assignments for the committee's final report.

APPENDIX C

Meeting Summary

Meeting of the Committee on Electrometallurgical Techniques for DOE Spent Fuel Treatment
J. Erik Jonsson Woods Hole Center
of the National Academies
September 19-21, 1999

SEPTEMBER 19, 1999—AGENDA

2:00 p.m.	**Closed Session**—Preliminary Discussion of Meeting Goals
2:30 p.m.	**Open Session**—Update on Argonne National Laboratory's Electrometallurgical Demonstration Project—Robert Benedict, Argonne National Laboratory
4:00 p.m.	**Closed Session**—Discussion of Final ANL Results and Preliminary Discussion of the Committee's Final Report
5:00 p.m.	Adjourn

SEPTEMBER 20, 1999—AGENDA

7:45 a.m.	**Closed Session**—Discussion of the Final Report Draft
8:00 a.m.	Chapter 1—Introduction
8:30 a.m.	Chapter 2—Historical Development
9:00 a.m.	Chapter 3—The Electrometallurgical Technique at ANL
9:45 a.m.	Product Streams Produced by the Electrometallurgical Process
10:15 a.m.	Break
10:30 a.m.	Post-demonstration Activities
11:00 a.m.	Previous Committee Recommendations
11:30 a.m.	Discussion of the Draft
12:00 p.m.	Lunch
1:00 p.m.	**Open Session**—Questions for ANL Personnel
2:00 p.m.	**Closed Session**—Writing Breakout Session
4:00 p.m.	Discussion of the Updated Report Draft

4:30 p.m.   Preliminary Discussion of Findings and Recommendations
5:00 p.m.   Adjourn

SEPTEMBER 21, 1999—AGENDA

7:45 a.m.   **Closed Session**—Findings and Recommendations for the Final Report
10:00 a.m. Break
10:15 a.m. Final Agreement on Findings and Recommendations
11:00 a.m. Outstanding Issues—Post Meeting Requirements, Planning Schedule for Completion and Release of the Committee's Final Report
12:00 p.m. Adjourn

SUMMARY OF PRESENTATIONS

Robert W. Benedict, ANL, spoke to the committee about the spent fuel demonstration project status. He gave the following summary of the repeatability results for driver electrorefining. The specific success criterion goal was to freeze process modifications and operating parameters while demonstrating a continuous throughput of 16 kg of driver uranium per month over a 3-month period. The repeatability demonstration began on November 14, 1998, and ended on January 22, 1999 (61 working days). The average treatment rate was approximately 24 kg per month.

For the blanket throughput, demonstration results were as follows. The specific success criterion goal was to treat 150 kg of blanket uranium in 1 month. The 1-month demonstration began on July 17, 1999, and ran through August 15, 1999 (30 days). The unit throughput for the blanket chopper was 164 kg; for the Mark-V electrorefiner, it was 205 kg; for the cathode processor it was 206 kg; and for the casting furnace it was 177 kg.

Mark-V electrorefiner process improvements since the completion of the throughput demonstration include the following. Stall recovery software has been implemented. The software automatically restarts the anode rotation after a stall. Using the demonstration operating conditions, the average production rate increased from 212 g of uranium per hour to 260 g of uranium per hour. Anode agitation software has been implemented along with stall recovery software. The software rotates that anode 90 to 180 degrees forward, then 45 to 90 degrees backward continuously. This greatly reduces the tendency to stall (0-4 stalls per product collector versus 16-20 stalls in the demo mode). The average production rate was increased to 350-400 g of uranium per hour.

Significant achievements for the Mark-V electrorefiner include the following. The latest run allows 400 g of uranium per hour per ACM as the average production rate, with 60 percent equipment utilization per ACM. The four ports are operational. Operation of the four ports is simultaneous, with routine operation of two ports possible. More than 13 blanket assemblies have been treated. Control software allows unattended operation. Two product collector harvesting methods have been developed: a bake-out oven with a gravity-assisted product dump at 500 °C, and a product collector harvesting tool that has a rotating multibladed tool used for grinding out product at room temperature.

The cathode processor treated 40 driver batches, 14 blanket batches, and 8 cladding hull batches. The casting furnace treated 40 driver batches, 14 blanket batches, and 7 metal waste batches.

For the driver fuel, the cathode processor/casting operating conditions were as follows. The cathode processor has a maximum crucible temperature of 1200 °C. The operating pressure was 0.1 torr, with isolation for cadmium. The salt distillation step took place over 1 hour at 1100 °C. Casting took place with a maximum crucible temperature of 1300 °C. The operating pressure was 900 torr until cast, and there was one stir cycle.

Significant accomplishments in the treatment process include the following. Driver treatment has processed 100 driver assemblies in 3 years. Eight assemblies were treated in 1 month. One thousand one hundred and ten kilograms of low-enriched uranium were cast. The cathode processor batch size increased from 12 to 19 kg. The casting furnace batch size increased from 36 to 54 kg. Blanket treatment has processed 13 of 25 blanket assemblies. The Mark-V electrorefiner has run 21 batches of irradiated blankets. Three hundred fifty five kilograms of blanket product have been cast. The blanket element chopper is operational.

Results for metal waste casting include the following. Three MWF ingots in the FCF that each accommodate an amount of cladding hulls corresponding to two driver fuel assemblies have been cast. The third MWF ingot with blanket cladding hulls substituted for driver hulls has been cast and characterized. The conclusion was that the appropriate casting conditions and waste characteristics have been determined and implemented for both driver and blanket hulls.

For the CWF, production of demonstration-scale irradiated samples involved the following. Irradiated salt was removed from the electrorefiner and transferred to the HFEF. Mill/classifier tests were performed to confirm the behavior of electrorefiner salt. A batch was successfully processed through the V-mixer. Ten radioactive samples and witness tubes have been processed through the HIP.

Significant accomplishments in waste activities include the following. A stainless steel-zirconium alloy continues as the MWF. The test matrix for qualification testing has been established. Three of three full batches of irradiated cladding hulls have been cast. Spiked and cold sample castings are complete. Waste qualification testing had started by the time of the meeting (July 21, 1999). Glass-bonded sodalite is the ceramic waste form. Initial uranium and plutonium behavior studies are available. Nonradioactive demonstration-scale equipment testing is complete. Equipment has been installed in the HFEF. Laboratory-scale samples containing plutonium for accelerated alpha decay tests have been fabricated. All 10 demonstration scale-cans were processed.

For the Environmental Impact Statement for the treatment and management of sodium-bonded spent nuclear fuel, the following schedule was presented. The notice of intent was published in February 1999. Scoping meetings were held in March 1999. The draft document was available in July 1999. Public hearings took place in August 1999. The final document was scheduled to be ready in January 2000, with the record of decision to be issued in February 2000.

A number of specific reports for the demonstration evaluation were issued. Overall demonstration reports include the following:

- *Spent Fuel Treatment Demonstration Final Report*,[1]
- *Production Operations for the Electrometallurgical Treatment of Sodium-bonded Spent Nuclear Fuel*,[2]
- *Analysis of Spent Fuel Treatment Demonstration Operations*,[3] and
- *Uranium Disposition Options*.[4]

Treatment operation reports include the overall treatment report:
- *Development of Cathode Processor and Casting Furnace Operating Conditions*.[5]

Driver treatment reports include:
- *Process Description for Driver Fuel Treatment Operation*,[6] and
- *Development of the Electrorefining Process for Driver Fuel*.[7]

---

[1]R.W. Benedict, H.F. McFarlane, S.P. Henslee, M.J. Lineberry, D.P. Abraham, J.P. Ackerman, R.K. Ahluwalia, H.E. Garcia, E.C. Gay, K.M. Goff, S.G. Johnsm, R.D. Mariani, S. McDeavitt, C. Pereira, P.D. Roach, S.R. Sherman, B.R. Westphal, R.A. Wigeland, and J.L. Willit, *Spent Fuel Treatment Demonstration Final Report*, NT Technical Memorandum No. 106, Argonne National Laboratory, Argonne, IL, 1999.

[2]K.M. Goff, L.L. Briggs, R.W. Benedict, J.R. Liaw, M.F. Simpson, E.E. Feldman, R.A. Uras, H.E. Bliss, A.M. Yacout, D.D. Keiser, K.C. Marsden, and C. Nielsen, *Production Operations for the Electrometallurgical Treatment of Sodium-Bonded Spent Nuclear Fuel*, NT Technical Memorandum No. 107, Argonne National Laboratory, Argonne, IL, 1999.

[3]H.E. Garcia, C.H. Adams, D.B. Barber, R.G. Bucher, I. Charak, R.J. Forrester, S.J. Grammel, R.P. Grant, R.J. Page, D.Y. Pan, A.M. Yacout, L.L. Burke, and K.M. Goff, *Analysis of Spent Fuel Treatment Demonstration Operations*, NT Technical Memorandum No. 108, Argonne National Laboratory, Argonne, IL, 1999.

[4]H.F. McFarlane, K.M. Goff, T.J. Battisti, B.R. Westphal, and R.D. Mariani, *Options for the Disposition of Uranium Recovered from Electrometallurgical Treatment of Sodium-Bonded Spent Nuclear Fuel*, NT Technical Memorandum No. 109, Argonne National Laboratory, Argonne, IL, 1999.

[5]B.R. Westphal, A.R. Brunsvold, P.D. Roach, K.C. Marsden, B.A. Jensen, and D.V. Laug, *Development of Cathode Processor and Casting Furnace Operating Conditions*, NT Technical Memorandum No. 110, Argonne National Laboratory, Argonne, IL, 1999.

[6]D. Vaden, B.R. Westphal, D.V. Laug, S.S. Cunningham, S.X. Li, T. A. Johnson, J.R. Krsul, and M.J. Lambregts, *Process Description for Driver Fuel Treatment Operations*, NT Technical Memorandum No. 111, Argonne National Laboratory, Argonne, IL, 1999.

[7]E.C. Gay, S.X. Li, R.K. Ahluwalia, D. Vaden, S.R. Sherman, and M.A. Power, *Development of the Electrorefining Process for Driver Fuel*, NT Technical Memorandum No. 112, Argonne National Laboratory, Argonne, IL, 1999.

Blanket treatment reports include:
- *Process Description for Blanket Treatment Operations*,[8] and
- *Development of the Electrorefining Process for Blanket Fuel*.[9]

For waste operations and qualification, the following reports have been issued. Overall waste reports include:
- *Waste Form Qualification Strategy*,[10]
- *Waste Acceptance Product Specifications*,[11]
- *Waste Compliance Plan*,[12]
- *Waste Form Degradation*,[13] and
- *Waste Form Degradation and Repository Performance Modeling*.[14]

Ceramic waste reports include:
- *Ceramic Waste Form Process Qualification Plan*,[15] and
- *Ceramic Waste Form Handbook*.[16]

Metal waste reports include
- *Metal Waste Form Process Qualification Plan*,[17] and
- *Metal Waste Form Handbook*.[18]

In summary, 100 drivers were treated by June 1999. Blanket treatment of 150 kg of uranium per month was successfully completed in August 1999. Thirteen blankets had been treated by the time of the meeting (September 19, 1999). Ten radioactive demonstration ceramic waste cans were produced in the summer of 1999. The Environmental Impact Statement was in the public comment period at the time of the meeting.

---

[8] S.R. Sherman, D. Vaden, R.D. Mariani, B.R. Westphal, T.S. Bakes, S.S. Cunningham, B.A. Jensen, T.A. Johnson, D.V. Laug, and J.R. Krsul, *Process Description for Blanket Fuel Treatment Operations*, NT Technical Memorandum No. 113, Argonne National Laboratory, Argonne, IL, 1999.

[9] E.C. Gay, S.R. Sherman, J.L. Willit, and R.K. Ahluwalia, *Development of the Electrorefining Process for Blanket Fuel*, NT Technical Memorandum No. 114, Argonne National Laboratory, Argonne, IL, 1999.

[10] T.P. O'Holleran, R.W. Benedict, and S.G. Johnson, *Waste Form Qualification Strategy for the Metal and Ceramic Waste Forms from Electrometallurgical Treatment of Spent Nuclear Fuel*, NT Technical Memorandum No. 115, Argonne National Laboratory, Argonne, IL, 1999.

[11] T.P. O'Holleran, D.P. Abraham, J.P. Ackerman, K.M. Goff, S.G. Johnson, and D.D. Keiser, *Waste Acceptance Product Specifications for the Waste Forms from Electrometallurgical Treatment of Spent Nuclear Fuel*, NT Technical Memorandum No. 116, Argonne National Laboratory, Argonne, IL, 1999.

[12] Argonne National Laboratory-West, *Waste Form Compliance Plan for the Waste Forms from Electrometallurgical Treatment of Spent Nuclear Fuel*, W0000-0062-ES, Rev. 00, Argonne National Laboratory, Idaho Falls, ID, 1999.

[13] R.A. Wigeland, L.L. Briggs. T.H. Fanning, E.E. Feldman, E.E. Morris, and M.C. Petri *Waste Form Degradation and Repository Performance Modeling*, NT Technical Memorandum No. 117, Argonne National Laboratory, Argonne, IL, 1999.

[14] R.A. Wigeland, L.L. Briggs, T.H. Fanning, E.E. Feldman, E.E. Morris, and M.C. Petri, *Waste Form Degradation and Repository Performance Modeling*, NT Technical Memorandum No. 117, Argonne National Laboratory, Argonne, IL, 1999.

[15] K.M. Goff, J.P. Ackerman, M.F. Simpson, M.C. Hash, K.J. Bateman, T.J. Battisti, and K.L. Hirsche, *Ceramic Waste Form Process Qualification Plan*, NT Technical Memorandum No. 118, Argonne National Laboratory, Argonne, IL, 1999.

[16] W.L. Ebert, D.W. Esh, S.M. Frank, K.M. Goff, M.C. Hash, S.G. Johnson, M.A. Lewis, L.R. Morss, T.L. Moschetti, T.P. O'Holleran, M.K. Richman, W.P. Riley, Jr., L.J. Simpson, W. Sinkler, M.L. Stanley, C.D. Tatko, D.J. Wronkiewicz, J.P. Ackerman, K.A. Arbesman, K.J. Bateman, T.J. Battisti, D.G. Cummings, T. DiSanto, M.L. Gougar, K.L. Hirsche, S.E. Kaps, L. Leibowitz, J.S. Luo, M. Noy, H. Retzer, M.F. Simpson, D. Sun, A.R. Warren, and V.N. Zyryznov, *Ceramic Waste Form Handbook*, NT Technical Memorandum No. 119, Argonne National Laboratory, Argonne, IL, 1999.

[17] B.R. Westphal, K.C. Marsden, S.M. McDeavitt, D.D. Keiser, Jr., D.P. Abraham, R.H. Rigg, B.A. Jensen, and D.V. Laug, *Metal Waste Form Process Qualification Plan*, NT Technical Memorandum No. 120, Argonne National Laboratory, Argonne, IL, 1999.

[18] D.P. Abraham, S.M. McDeavitt, D.D. Keiser, S.G. Johnson, M.L. Adamic, S.A. Barker, T.D. DiSanto, S.M. Frank, J.R. Krsul, M. Noy, J.W. Richardson, Jr., and B.R. Westphal, *Metal Waste Form Handbook*, NT Technical Memorandum No. 121, Argonne National Laboratory, Argonne, IL, 1999.

APPENDIX D

Recommendations and Selected Findings and Conclusions from Previous Reports of the Committee on Electrometallurgical Techniques for DOE Spent Fuel Treatment

PHASE ONE

A Preliminary Assessment of the Promise of Continued R&D into an Electrometallurgical Approach for Treating DOE Spent Fuel (Report 1, 1995)

The committee's first report was issued shortly after the committee began its work. Following briefings from a number of individuals in the nuclear energy field concerning different aspects of the electrometallurgical process, the committee made the following two statements:

- The committee concluded that electrometallurgical techniques represent a sufficiently promising technology for treating a variety of DOE spent fuels to warrant continued R&D in federal FY96.
- During the next 12 months the DOE should closely follow the progress of the proposed R&D program to determine whether it should be continued beyond FY96.

An Assessment of Continued R&D into an Electrometallurgical Approach for Treating DOE Spent Nuclear Fuel (Report 2, 1995)

As the demonstration project progressed, the committee recommended three criteria for judging the success of the demonstration project (a minimum definition of "successful application"):

- Demonstration of batch operation of an electrorefiner and a cathode processor with a capacity of approximately 200 kg/day of radioactive EBR-II spent fuel without failure for about 30 days.
- Quantification (for both composition and mass) of recycle, waste, and product streams that demonstrate projected material balance with no significant deviations.
- Demonstration of an overall dependable and predictable process, considering uptime, repair and maintenance, and operability of linked process steps.
- Demonstration that releases of radioactivity remain at or below those levels anticipated and specified in equipment design and operating plans. Exposure of operating personnel to radiation must be minimal and must in all cases remain below limits set by the U.S. Nuclear Regulatory Commission.

The committee then proposed that the demonstration project should proceed. Its recommendation included options for the treatment of other spent fuel in the DOE inventory:

ANL should proceed with its development plan in support of the EBR-II demonstration. Further development of the lithium reduction process should be carried out only if the DOE decides that it is likely that the electrometallurgical approach will be pursued as a possible treatment for the oxidized N-reactor fuel at Hanford. If the EBR-II demonstration is not accomplished successfully, the ANL program on electrometallurgical processing should be terminated. On the other hand, if the EBR-II demonstration is successful, the DOE should revisit the ANL program at that time in the context of a larger, "global" waste management plan to make a determination for possible continuance.

PHASE TWO

An Evaluation of the Electrometallurgical Approach for Treatment of Excess Weapons Plutonium (Report 3, 1996)

As a result of its statement of task for phase two (Appendix A), the committee's composition was changed to emphasize expertise in weapons plutonium. The committee's third report addressed the issue of the use of electrometallurgical technology for treatment of excess weapons plutonium:

- Modified Spent Fuel Processing Flow Sheet Recommendations:
  — Pretreatment requirements for the nonmetal plutonium feed streams should be determined and, if possible, R&D should be started to validate the treatment and subsequent compatibility with the electrometallurgical process.
  — The effects of major impurities such as additional salts (NaCl/KCl and $CaCl_2$) and other impurities such as Si, Mg, and C on the performance of the electrometallurgical treatment operations should be evaluated.
- Greater priority should be given to the development of a strategy and a relevant test protocol to demonstrate acceptability of waste forms. This activity is of the highest importance relative to all other aspects in the development of the electrometallurgical technique for WPu disposition.
- A decision on the use of the electrometallurgical technique for weapons plutonium disposition cannot be made until the demonstration of this technology shows whether or not this process is viable for treating DOE spent fuels. If a weapons plutonium disposition technology is to be selected for use with weapons pits before the electrometallurgical technology demonstration program is concluded, this committee recommends that the electrometallurgical technique not be included as a candidate technology.
- The potential of the electrometallurgical technique as an adjunct for long-term disposition of non-pit excess plutonium remains a possibility, but the technology is still at too early a stage of development to be evaluated relative to disposition alternatives such as glass or MOX.

Electrometallurgical Techniques for DOE Spent Fuel Treatment: A Status Report on Argonne National Laboratory's R&D Activity (Report 4, 1996)

The committee's fourth report addressed the first task for phase two. The report provided an ongoing evaluation of the scientific and technological aspects of the R&D program for spent fuel treatment. The findings and recommendations relate to the demonstration project:

- The committee recommends that DOE assign high priority to authorization of hot operations at ANL-W.
- The committee recommends that ANL's ongoing studies be extended to include efforts aimed at defining

APPENDIX D

the phase changes in the salt-loaded zeolite during hot isostatic pressing and determining the fate of the salt, which would no longer be as well isolated from the environment.
- The committee recommends that attention be given to establishing the performance of both the zeolite and metal waste forms under conditions relevant to their disposal in a geological repository.
- The committee recommends that the several aspects of ANL's substantial effort in waste form development be integrated into a formal, comprehensive program plan.
- The committee also recommends that ANL establish a program of regular, formal meetings between ANL personnel and staff of the DOE's Yucca Mountain project as a useful (and perhaps essential) step in guiding ANL's future testing program.
- The committee recommends that upon satisfactory completion of the demonstration with EBR-II fuel, the electrometallurgical technique should be evaluated in the broader context of alternative technologies for processing spent nuclear fuel.

Electrometallurgical Techniques for DOE Spent Fuel Treatment: Fall 1996 Status Report on Argonne National Laboratory's R&D Activity (Report 5, 1997)

The committee's fifth report continued to fulfill the first task for phase two by providing an ongoing evaluation of ANL's R&D activity. The recommendations in this report cover most aspects of ANL's demonstration project.

Spent Fuel Recommendations

- A well-defined set of performance criteria needs to be developed. The criteria would provide ANL with a clear set of objectives. The achievement of those objectives would better position ANL to request approval to proceed to additional applications of its electrometallurgical technology program.
- A more focused, better-coordinated testing and implementation plan is needed between ANL-E and ANL-W to ensure that performance criteria and demonstration schedules are met.

Waste Form Recommendations

ANL should develop and implement immediately an overall strategic plan that defines the following:
- The planned state of waste form development at the end of the demonstration phase and the objectives that will remain to be addressed and
- The methods for ensuring optimal, synergistic use of all ANL resources for ceramic waste form development and evaluation.

Electrometallurgical Techniques for DOE Spent Fuel Treatment: Status Report on Argonne National Laboratory's R&D Activity Through Spring 1997 (Report 6, 1997)

Like report 5, report 6 continued to address the first task for phase two. This report reiterated previous committee recommendations:

- The committee reaffirms its overall recommendation of the July 1995 report [Report 2]. The committee encourages ANL to proceed aggressively to resolve the R&D issues and move rapidly into a demonstration phase that identifies process definitions and conditions.

The committee went on to present new recommendations related to the demonstration project.

- The committee looks forward to receiving the demonstration project implementation plan after it is approved by DOE.
- Before the demonstration is completed, DOE should establish criteria for success in the demonstration phase to allow evaluation of the electrometallurgical technology for further use in treating DOE spent fuel.
- A new EA will be required before additional EBR-II spent fuel can be treated. DOE should begin plans for such an EA now so that its preparation does not become the source of a major operational delay, if the current demonstration project is successful.
- The committee continues to believe that successful demonstration of the electrometallurgical process for treating EBR-II fuel is essential to support development of applications of this technique to treatment of other DOE spent fuels. ANL's research efforts have involved the investigation of the electrometallurgical technology for treatment of non EBR-II fuels such as the N-reactor fuel. However, the DOE Office of Environmental Management (EM) may proceed with plans for the N-reactor fuel that do not include the use of electrometallurgical technology. Since the current approach of DOE-EM is to develop project plans for implementation within the next 10 years, the offices of Nuclear Energy (NE) (which funds the present program) and Environmental Management (EM) should maintain close contact to ensure proper coordination of their activities.
- DOE should establish acceptance criteria for waste forms scheduled for storage in a geological repository.
- The committee suggests that ANL utilize external technical experts in specific scientific areas of the program. These technical experts should be recognized for their in-depth knowledge in particular technical areas.

PHASE THREE

Electrometallurgical Techniques for DOE Spent Fuel Treatment: Spring 1998
Status Report on Argonne National Laboratory's R&D Activity (Report 7, 1998)

The committee's seventh report concentrated on the second and third tasks for the third phase. The committee examined electrometallurgical technology in light of other possible treatment technologies for spent nuclear fuel, and it evaluated the success criteria developed by ANL and DOE for the demonstration project:

- Confirmation that the waste forms produced by EMT are acceptable within the DOE's Office of Radioactive Waste (DOE-RW) Office of Civilian Radioactive Waste Management (OCRWM) program for final geological disposal must be a key component in a full qualification of the EMT process.
- If DOE concludes that the EMT process is unsuitable for processing the remainder of EBR-II fuel, then the PUREX process could be evaluated for its applicability to treatment of EBR-II fuel. However, a significant issue for treating EBR-II fuel at SRS by PUREX relates to public concerns about the transportation of the fuel from the current storage site at ANL-W to SRS.
- In recognition of the progress that ANL has made in the demonstration and in accord with the committee's previous recommendations, the committee recommends to DOE that ANL's demonstration project be carried to completion.
- The committee finds that the criteria established by U.S. DOE are reasonable for judging the success of the EBR-II spent fuel treatment demonstration.

APPENDIX D

Electrometallurgical Techniques for DOE Spent Fuel Treatment: Status Report on
Argonne National Laboratory's R&D Activity as of Fall 1998 (Report 9, 1999)

The committee's eighth report fulfilled its task to provide an ongoing evaluation of ANL's demonstration project:

- ANL should broaden its perspective regarding the Mark-V ACM by seeking information about the following:
  — The physical, morphological, and mechanical characteristics (e.g., plasticity) of the uranium/salt mixture produced during HTER operating conditions;
  — The electrochemical behavior of uranium in molten LiCl-KCl under HTER operating conditions; and
  — Useful strategies from the metal electrowinning industry that can be applied to the uranium electrometallurgical process.
- ANL should evaluate the potential impact of the higher salt content of the Mark-V HTER product on the performance of the cathode processor.
- ANL should evaluate the effects of cathode surface roughness on the adhesion of the uranium deposit; other materials or metallic coatings that might reduce adhesion of uranium on the ACM cathode should be considered.
- Surface analysis by XPS or AES should be performed for selected samples drawn from the characterization tests. The committee notes that ANL-E has one of the leading experts in this area. It is recommended that only a few of these samples be fully characterized.
- ANL needs to refocus near-term testing by reducing product consistency testing under relatively mild conditions and instead emphasizing product performance testing under more stringent conditions that may reveal significant corrosion effects and address success criterion 2, goal 2.
- ANL should evaluate how the MWF performance model will be used and whether ongoing MWF testing will provide information needed for developing the performance model.
- Assuming that an increased amount of glass in the waste form is acceptable, then in addition to HUP, conventional cold pressing and sintering should also be considered as a viable processing option.
- The committee believes that characterization of the ceramic waste form should be accelerated in order to determine the mechanism of transformation of salt-loaded zeolite 4A to sodalite.

Electrometallurgical Techniques for DOE Spent Fuel Treatment: An Assessment of
Waste Form Development and Characterization (Report 9, 1999)

In conversations between personnel of DOE and the NRC, it was agreed that the committee would produce a report that would examine the issue of waste form testing and development as it related to ANL's demonstration project. Specifically, the committee assessed whether the testing plan in place for the ceramic and metal waste forms was the proper one to qualify these waste forms for placement in a geologic repository. Although final criteria for repository placement have not yet been finalized, the committee based its assessment on the question of whether it believed that the testing program was valid. Relevant findings and conclusions from this report are included with the recommendations:

Finding: From interactions with the DOE's Office of Civilian Radioactive Waste Management (RW), ANL has developed a strategy appropriately based on RW's waste acceptance criteria for the characterization of its MWF and CWF for eventual acceptance by RW for geologic storage and/or disposal based on RW.

Conclusion: Continued interaction between ANL and RW will become even more important in the post-demonstration phase.

Conclusion: There remains uncertainty regarding which DOE organization will be charged with the ultimate responsibility for performance-confirmation testing of waste forms suitable to support a repository licensing decision. As this uncertainty in responsibility could lead to costly duplication of effort and lack of consensus among DOE organizations regarding data supporting future decisions, DOE should take the lead in achieving a documented resolution to this issue.

Finding: An SS-5 Zr MWF shows severe rusting and pitting in the vapor hydration test.

Conclusion: Data for the SS 15-Zr MWF standard need to be obtained.

Recommendation: Surface analysis by X-ray photoelectron spectroscopy (XPS) or Auger electron spectroscopy (AES) should be performed for selected samples to determine the chemical composition of passivation filings and/or corrosion products. Because a large number of samples to be tested differ only very slightly in minor alloying elements, it is recommended that only a few of these samples be subjected to full characterization. These samples should be selected using a statistical analysis approach.

Finding: Some of the corrosion products, which may sequester radionuclides, might remain on the sample surface and might not be detected by solution analysis.

Finding: Results from corrosion testing of the MWF in rather benign environments suggest that the corrosion behavior of the MWF is similar to that of stainless steel.

Finding: At the present time, ANL has not indicated how it plans to conduct crevice corrosion studies.

Finding: ANL has carried out a large number of corrosion tests in solutions which are not expected to lead to significant corrosion damage.

Recommendation: Instead of continuing to conduct a large number of such tests using mild conditions, it would be better to subject a few carefully selected samples to additional evaluation by surface analysis to determine the chemical composition of the corrosion products. It may be better to concentrate on a few key samples, expose them at higher temperatures, and then obtain electrochemical and surface analysis data.

Finding: ANL's tests over the several months duration of the test indicate that the CWF dissolves at a rate equal to or less than reference defense high level waste borosilicate glass.

Conclusion: If there is no change in long-term release mechanism under simulated repository conditions, the release performance of CWF (dissolution rate) will be at least comparable to borosilicate glass.

Finding: During the conduct of the alpha-decay tests, plutonium oxide was observed as nanocrystals in the grain boundaries.

Conclusion: Plutonium may not be in the sodalite phase. Its presence in potentially colloid-sized products may have implications on the long-term release behavior of plutonium and any other radionuclides that also segregate into such colloid-size phases.

APPENDIX D

Recommendation: The EMT Program should continue to evaluate and demonstrate that test protocols and conceptual models developed for monolithic single-phase borosilicate glass can adequately represent the behavior of the nonhomogeneous multiphase EMT CWF.

Finding: The Material Characterization Center Test (MMC-1) and Product Consistency Test (PCT) designed to model the release behavior of inert, major components of the CWF may be irrelevant with respect to evaluating the release of plutonium and other actinides partitioned into separate oxide phases.

Conclusion: The committee believes that ANL is taking appropriate steps to coordinate its waste-qualification program with the DOE RW repository program. It remains undemonstrated, however, that direct adaptation of test procedures and models developed for measuring the rate of general corrosion of the matrix of homogeneous, vitrified HLW forms are sufficient for evaluating the performance of the heterogeneous, crystalline CWF under long-term repository conditions.

Conclusion: These continuing concerns are not expected to jeopardize the timely completion of the EBR-II demonstration project in 1999, but attention should be devoted to their resolution prior to extending the EMT process past the demonstration.

Conclusion: Alternative, conservatively bounding strategies for assuring safe disposal of such relatively small quantities of novel HLW may result in significant cost avoidance while still protecting public safety.

APPENDIX E

Abbreviations and Acronyms

ACM	anode-cathode module
AES	Auger electron spectroscopy
ALMR/IFR	Advanced Liquid-Metal Reactor/Integral Fast Reactor
ANL	Argonne National Laboratory
ANL-E	Argonne National Laboratory-East (Argonne, Illinois)
ANL-W	Argonne National Laboratory-West (Idaho Falls, Idaho)
ANSTO	Australian Nuclear Science and Technology Organisation
ASTM	American Society for Testing and Materials
CCDF	compliance cumulative distribution functions
CISAC	Committee on International Security and Arms Control
CWF	ceramic waste form
DHLW	Defense Program High-Level Waste
DOE	U.S. Department of Energy
DOE-EM	U.S. Department of Energy's Office of Environmental Management
DOE-RW	U.S. Department of Energy, Office of Civilian Radioactive Waste Management
DSC	differential scanning calorimetry
DU	depleted uranium
DWPF	Defense Waste Processing Facility
EA	environmental assessment
EBR-II	Experimental Breeder Reactor-II
EBS	engineered barrier system
EIS	environmental impact statement
EMT	electrometallurgical technology
FCF	fuel conditioning facility
FDB	fuel dissolution basket
FFTF	Fast Flux Test Facility
GMODS	glass material oxidation and dissolution
HEU	highly enriched uranium
HFEF	Hot Fuel Examination Facility

HIP	hot isostatic press
HLW	high-level waste
HTER	high-throughput electrorefiner
HUP	hot uniaxial press
INEEL	Idaho National Engineering and Environmental Laboratory
INTEC	Idaho Nuclear Technology and Engineering Center
J-13	simulated Yucca Mountain well water
LEU	low enriched uranium
LLW	low-level waste
LWR	light water reactor
MCC	Material Characterization Center
MGDSWAC	mined geologic disposal system waste acceptance criteria
MOA	memorandum of agreement
MOX	mixed uranium-plutonium oxide fuel
MSRE	molten salt reactor experiment
MTHM	metric tons heavy metal
MWF	metal waste form
NE	U.S. Department of Energy Office of Nuclear Energy, Science, and Technology
PCT	product consistency test
R&D	research and development
RIP	repository integration program
SEM	scanning electron microscopy
SNF	spent nuclear fuel
SRS	Savannah River site
SS	stainless steel
TCLP	toxicity characteristic leaching procedure
TEM	transmission electron spectroscopy
TRU	transuranic elements
TSPA	total system performance assessment
WAPS	waste acceptance product specifications
WCP	waste form compliance plan
WIPP	waste isolation pilot plant
WISP	Waste Isolation System Panel
WPu	weapons plutonium
WQR	waste form qualification report
XAFS	X-ray absorption fine structure spectroscopy
XPS	X-ray photoelectron spectroscopy
XRD	X-ray diffraction